Tourism on the Verge

Series Editor

Ulrike Gretzel, Annenberg School for Communication and Journalism, University of Southern California, Los Angeles, CA, USA

Tourism on the Verge aims to provide a holistic understanding of various phenomena that shape tourism and hospitality in profound and lasting ways, approaching research topics and practical issues from multiple perspectives. Each volume in the series will address transformations within a particular area, in order to advance both our theoretical understanding and practical applications. Books should be conceptual in nature and make highly relevant contributions to explaining these phenomena. Attention should also be drawn to cutting-edge methods, in order to stimulate new directions in tourism research. The series will publish works of the highest quality and which follow a logical structure, rather than merely presenting a collection of articles loosely related to a topic. Book editors will be asked to write a strong introductory chapter that offers a comprehensive overview of the selected topic areas / fields. Presenting a unique blend of scholarly research, the series will be an unparalleled reference source.

Cecilia Pasquinelli • Mariapina Trunfio

Sustainability-oriented Innovation in Smart Tourism

Challenges and Pitfalls of Technology Deployment for Sustainable Destinations

Springer

Cecilia Pasquinelli
Department of Management
and Quantitative Studies
University of Naples Parthenope
Naples, Italy

Mariapina Trunfio
Department of Management
and Quantitative Studies
University of Naples Parthenope
Naples, Italy

ISSN 2366-2611 ISSN 2366-262X (electronic)
Tourism on the Verge
ISBN 978-3-031-33676-8 ISBN 978-3-031-33677-5 (eBook)
https://doi.org/10.1007/978-3-031-33677-5

This Springer imprint is published by the registered company Springer Nature Switzerland AG
The registered company address is: Gewerbestrasse 11, 6330 Cham, Switzerland

Contents

About the Authors

Cecilia Pasquinelli is an Associate Professor of Economics and Business Management at the Department of Management and Quantitative Studies, University of Naples Parthenope, Italy. She obtained her PhD in Management, Competitiveness and Development at Scuola Superiore Sant'Anna (Italy). She previously worked as a researcher in the GSSI Cities, Gran Sasso Science Institute in L'Aquila (Italy) and in the Department of Social and Economic Geography at Uppsala University (Sweden), where she joined the Centre for Research on Innovation and Industrial Dynamics. She is an Associate Editor of the Journal of Place Management and Development (Emerald). She is a Senior Fellow at the Institute of Place Management, Manchester Metropolitan University (UK) and an Affiliate Professor at the Institute of Management, Scuola Sant'Anna (Italy). She developed several action research projects on tourism development, local sustainable development and innovation policies in collaboration with local, regional and national authorities. She is a lecturer in Destination Management and Brand Management at Parthenope University, and she was a guest lecturer at several national and international universities. Her research interests include place branding and marketing, sustainable tourism and urban tourism management, smart tourism, place of origin and local development. Her work was published in various international peer-reviewed journals, including the Journal of Sustainable Tourism, Urban Studies, Regional Studies, Cities, Marketing Theory, British Food Journal, Journal of Hospitality Marketing & Management and Environment Planning Studies. She co-edited the book *Tourism in the City: Towards an Integrative Agenda on Urban Tourism*, Springer. She developed experience as a consultant in place marketing and foreign direct investment promotion, working at OCO Global Ltd. in Paris (France).

Mariapina Trunfio (PhD) is an Associate Professor of Economics and Business Management and Director of the Master in Tourism & Hospitality Management at the Department of Management and Quantitative Studies, University of Naples Parthenope, Italy. She teaches management of tourism enterprises and international service management. She was the Scientific Supervisor of several innovative tourism research projects and tourism strategic plans, including the Italian National Tourism Strategic Plan. She was a Member of diverse scientific advisory boards in tourism and hospitality innovation, including the Advisory Boards of BTO (Be Travel Onlife)-Digital Strategy and Dolomiti Paganella Future Lab. Her current research focuses on tourism innovation, smart destination, digital and social media marketing, augmented reality, virtual reality and mixed reality in tourism and cultural heritage, metaverse, destination management and governance, local development, overtourism and sustainable development, cultural heritage and creativity, cultural diversity and entrepreneurship. She extensively published over 100 in leading journals, monographs, book chapters and conference papers in these areas. Her papers were published in several international peer-reviewed journals, including the Information & Management, Current Issue in Tuorism, Journal of Destination Marketing & Management, International Journal of Contemporary Hospitality Management, Journal of Sustainable Tourism, Journal of Hospitality & Tourism Research, Cities and British Food Journal. She received several international best papers awards.

List of Figures

List of Tables

Chapter 1
Sustainability-Oriented Innovation in Smart Destinations: An Integrative Perspective

1.1 Rethinking Tourism Innovation in the Tech-Clash Era: Towards a Smart Tourism Mindset for Sustainability and Human Wellbeing

The increasing attention to the disruptive power of smart technologies has fascinated academics, practitioners, industries, and policymakers worldwide in the last two decades. Mythologising technologies, practitioners, and policymakers exploited several ICT tools to solve business challenges and societal imbalances. Leveraging technology-driven innovation has been seen as an all-encompassing approach, underestimating the pitfalls and failures of a myopic technology-isolated vision.

Overcoming the roadmaps of the past two decades, unconditional trust for new technologies is reducing, and consolidated approaches and business models of technology exploitation are in crisis. Academics, practitioners, and policymakers have wondered why digital strategies and technological investments fail and which are the main drivers to cope with technology-driven innovation limitations and pitfalls. They are called to recognise today's "tech-clash"—the conflict between stakeholders, business, and technology—and develop "provocative thinking, transformative insights, tangible outcomes" (Accenture 2020).

The "tech-clash" of the socio-economic scenario represents the starting point for this book. It aims to cast a critical light on the consolidated views of disruptive technological exploitation, laying the foundation stones of the academic debate on sustainability-oriented innovation in smart tourism destinations. Although the ongoing technology debate embraced tourism, research on innovation, sustainability, and technology remains separated, fragmented, and mainly conceptual. An integrative conceptual framework was proposed, interpreting "destination innovation as a complex and evolutionary knowledge-driven phenomenon resulting from human (organisational) technology interactions" (Trunfio and Campana 2019, p. 9).

Disruptive technologies innovation has been under observation, opening up spaces for rethinking consolidated models and conceptualising technological

C. Pasquinelli, M. Trunfio, *Sustainability-oriented Innovation in Smart Tourism*, Tourism on the Verge, https://doi.org/10.1007/978-3-031-33677-5_1

advances for solving societal imbalances and improving wellbeing. The literature has criticised the myopic vision of disruptive technological power, which underestimates human and social perspectives (Stankov and Gretzel 2020; Pencarelli 2020; Pasquinelli et al. 2023b; Trunfio et al. 2022). There is an urgent need to frame new routes of academic investigation, combining empirical analyses and theoretical advancements, to move technological power towards enhancing responsible behaviours, inclusiveness, and human wellbeing. Responsible and social innovations achieving sustainable and inclusive development are in focus in the academic debate (Edwards-Schachter 2018; Fagerberg 2018; Trunfio and Campana 2020; van der Have and Rubalcaba 2016). These embrace the notion of changing mindsets and creating hybrid innovation as mixed modes, recombining technology, society, economy, and sustainability (Kahn 2018; OECD 2017).

A human-centred perspective was introduced in the tourism debate underlining the urgent need to overcome the myopic technological vision. The academic investigation started to include minds, bodies, and wellbeing in travel, introducing a multi-stakeholder focus on hedonistic, altruistic, and meaningful experiences (Stankov and Gretzel 2020). The rhetoric of smart tourism has been under scrutiny, calling for the introduction of "liquidity into the smart tourism ecosystem" (Gretzel 2022) and investigating the limitations and risks of digital transformation (Stankov and Gretzel 2020; Trunfio and Campana 2019). Research started to investigate the dark side of human–technology interactions revealing limitations, barriers, and constraints in adopting and using technology as a substitute for or complement to the human touch (Bach and Scapin 2004; Bekele et al. 2018; Flavián et al. 2019; Trunfio et al. 2022). Gretzel (2021) called for the constructive and collective re-imagining of innovation by reframing the smart tourism mindset, shaping stakeholder ideas and values, and allowing a nurturing attitude to anticipate, drive, and implement technological change and pervasive innovation. This opens up new cognitive spaces for the future of the destination.

A paradigm shift based on a smart tourism mindset is needed, which engages the stakeholder in a co-evolutionary process of destination innovation, building on the broader vision of smartness in tourism, which sheds light on the combination between technology, business, and society. Although several papers called for research on the human side of digital exploitation, their contribution remained partial and fragmented, devoted to specific aspects of technological tools and applications. Rethinking innovation in the broader scenario of human–technology interaction represents a stimulating challenge for academics who are in charge to reframe unchartered territories and face up to conceptual pitfalls and technological limits.

1.2 Sustainability-Oriented Innovation in Smart Tourism: A Call for an Integrative Research

This book cross-fertilises existing theoretical domains of innovation, smart technologies, and sustainable tourism, introducing fresh knowledge into the academic debate and drawing implications for tourism businesses and policymakers. It proposes a conceptual advancement on sustainability-oriented innovation in the contemporary complexity of knowledge-based destinations, by investigating and discussing emerging smart destination models and technological solutions involving multiple stakeholders in sustainable development. This research combines research on technology- and social-driven innovation in tourism destinations and sustainable destination management research, providing an integrative understanding of the nexus between smartness and sustainability within destinations.

Although innovation and technology are widely considered the way to achieve sustainable tourism development, research that critically engages with the links between sustainability and smart technologies remains limited. Sustainability is often framed as an implication of the smart destination (Gretzel 2022; Shafiee et al. 2019; Trunfio and Pasquinelli 2021), with limited consolidation of a dedicated research agenda on the smart and sustainable destination in the academic debate. This contrasts with the significant emphasis on the smart and sustainable destination in the policy domain at local, national, and supranational levels.

This book develops a conceptual and empirical exploration to frame technology deployment as a mechanism for a transformative process leading the destination system to rebalance tourism disequilibria and effectively address tourism impacts and imbalances. Sustainability-oriented innovation comes from the broad field of inquiry on business innovation. It focuses on organisations' ability to change and reshape products, practices, and internal processes to boost social and environmental value creation (Adams et al. 2016). The technical and technological focus has dominated the sustainability-oriented innovation debate in the business context (Adams et al. 2016), but stakeholders' mindsets and cultural meanings were said to play a relevant role in such a process (Paramanathan et al. 2004; Adams et al. 2016; Pasquinelli et al. 2023a). Transformative effects were highlighted and discussed in industries, business management, and consumption behaviour (Coenen and Truffer 2012).

That is, the pursuit of sustainability brings about transformation processes in which technical and technological assets, mindsets and cultural assets, practices, and relations need to assist change and, consequently, evolve even more radically. From these premises, this book contributes to framing sustainability-oriented innovation as a technology-based process to achieve social, economic, and environmental value for the destination stakeholders. The book combines conceptual research and empirical analysis to frame an integrative agenda that outlines sustainability as an innovation process oriented to achieving social, economic, and environmental values for the destination stakeholders. It engages theoretically and empirically with the research question of how and to what extent smart technologies drive tourism innovation that is focused on sustainability. This book's core is analysing the modalities, challenges, opportunities, and pitfalls of technology deployment in rebalancing tourism disequilibria and pursuing sustainable and inclusive development. It allows us to outline scenarios of sustainability-oriented innovation, encompassing and intertwining technological, social, human, and cultural co-drivers as the necessary conditions for smart technologies to support sustainable destinations effectively.

1.3 The Pre-pandemic Legacy of the Overtourism Debate

The first wave of sustainable-oriented innovation can be positioned in the evolutionary process of the overtourism debate (Pasquinelli and Trunfio 2020a). The overtourism phenomenon in the pre-pandemic largely reframed the broader and long-standing sustainable tourism agenda by focusing on the social and environmental downsides of tourism intensification in the 2010s (Peeters et al. 2018; Milano et al. 2019; Séraphin et al. 2020; Pasquinelli and Trunfio 2020a, b; Koens et al. 2018).

Characterising, especially urban contexts but also non-urban regions, overtourism catalysed the academic, policymaker, and media attention for its capacity to reshape the economic, physical, and social landscapes of cities and regions (Peeters et al. 2018; Koens et al. 2018; Séraphin et al. 2020; Pasquinelli and Trunfio 2020a). Much attention was also catalysed by the emergence of tourism-phobia and anti-tourism movements and several protests against forms of tourism saturation (Colomb and Novy 2016; Milano 2017; Koens et al. 2018; Martín et al. 2018; Séraphin et al. 2018).

Scholars focused on the causes and impacts of overtourism, whilst solutions, tools, and approaches to addressing overtourism remained marginal to the academic debate (Pasquinelli and Trunfio 2020a). This tendency can be interpreted in line with the evident weaknesses of the sustainable tourism agenda to turn theoretical contributions into policy implications supporting concrete actions (Reinhold et al. 2015). Research is needed to frame solutions, actions, and tools into theoretical frameworks for interpreting their nature and implications (Koens et al. 2018; Pasquinelli and Trunfio 2020b). This need emerged from cities such as, for instance, Barcelona, Berlin, Venice, Florence, and Amsterdam that, being the international icons of the overtourism phenomenon (Milano 2017; Minoia 2017; Fava and Rubio 2017; Pasquinelli et al. 2022), promoted agreements and programmes to counteract the negative tourism impacts. Scholars are in charge of providing analytical support to this process.

In this frame, smart technologies to address overtourism and drive sustainable development have attracted some academic attention (Trunfio and Pasquinelli 2021). If ICT tools are often adopted as "a practical approach" for destination managers to mitigate negative tourism impacts (Ali and Frew 2014), the multifaceted concept of sustainability calls for deeper engagement with the analysis of smart technologies' roles alongside sustainable destination development, which represents a field of largely underdeveloped inquiry.

ICT tools were not significantly investigated in relation to their capacity to assist the destination's response to overtourism and the "overtourism hazard" (Muler Gonzalez et al. 2018; Pasquinelli and Trunfio 2020a, b, c). This contrasts with their evident potential in supporting key actions that were also discussed in the literature, such as optimisation strategies, visitor flow management, limitations to tourism loads in vulnerable places, and dispersal strategies (Dodds and Butler 2019; Oklevik et al. 2019; Séraphin et al. 2019; Trunfio and Pasquinelli 2021).

1.4 COVID-19 Pandemic: Continuities and Emerging Scenarios

Recent post-pandemic evolutions, following the global spread of the COVID-19 virus, further called attention to smart technologies as pragmatic answers to the several issues that tourism destinations have to address. Despite the different characteristics of pre- and post-pandemic tourism, sustainable destination development represents the common thread in the tourism debate. The COVID-19 pandemic created unexpected tourism scenarios and exerted transformative impacts on tourism systems, with structural consequences on the industry (Gossling et al. 2020). Studies proliferated and showed evidence of a change in tourism and in destinations, producing new frameworks to explain emerging aspects and trajectories characterising post-pandemic tourism phenomena (Sigala 2020).

Turnover and job losses in the hospitality sector, with related social costs, effects on vacation rental markets (UNWTO 2020; Dolnicar and Zare 2020; Karabulut et al. 2020; Sharma and Nicolau 2020), and, on the other hand, a change in tourists' travelling preferences, intentions, and behaviours (e.g. Del Chiappa et al. 2022; Matiza 2020; Li et al. 2021; Zhang et al. 2020, 2021; Hassan & Soliman, 2021; Talwar et al. 2022b) are the salient signs of potential tourism transformation.

Scholars have warned against the possibility that pre-pandemic imbalances may be firmly maintained, revealing a weak structural change following the COVID-19 crisis (Higgins-Desbiolles 2021; Collins-Kreiner and Ram 2020; Ioannides and Gyimothy 2020). Tourism studies should not overlook the significant links and continuities between pre- and post-pandemic tourism in terms of meaningful analytical dimensions, undertaken actions by key tourism stakeholders, and underlying tourism development models (Pasquinelli and Trunfio 2022). The acceleration of the adoption and acceptance of digital technologies (Gretzel et al. 2020) is certainly a dimension of the COVID-19 pandemic impact that should be considered. Physical mobility and travelling were not possible, and social distancing became an international standard during the lockdown phases, leaving room for interpersonal connections through digital tools, for leisure, working, shopping, and education. Virtual tourism and leisure became the only possibility to travel, with an increase in digital experience platforms (Wong et al. 2022; Zhang and Qiu 2022; Zhu and Cheng 2021; Pasquinelli et al. 2023b).

These premises suggest the importance of the pandemic crisis for shifting knowledge about digitalisation and sustainability (Talwar et al. 2022a) in the practice and in academic research. A systematic literature review focused on "sustainable innovation" and "sustainable digitalisation" in the hospitality and tourism industry during the pandemic (Elkhwesky et al. 2022). Product/service innovations, marketing innovations, organisational, and business model innovations were connected to the adoption of technology types.

This revealed that, since the beginning of the pandemic, hospitality, and the tourism industry relied significantly on network technologies (e.g. blockchains and social media) and data-processing technologies (for data analysis to assist decision-making and control). There was greater reliance on these than on physical–digital interface technologies (e.g. VR, AR, augmented virtuality, mixed realities, future realities, a real-time tour, and videoconferencing tools) and physical–digital process technologies (e.g. equipment and solutions for managing energy consumption and reducing emissions). These results represent a snapshot of the technologies emerging from the cases and practices analysed and discussed in the existing literature or based on conceptual research (24% of the reviewed studies were conceptual). Research is needed to deepen knowledge of the technological shift that might have emerged during the different pandemic phases.

An analysis of Twitter contents on Italian tourism in January 2021 highlighted the connection between digital technologies and the recovery-reform dichotomy in the COVID-19 pandemic debate on tourism (Pasquinelli and Trunfio 2022). ICT tools emerged as a topic, contributing to tourism innovation towards a necessary reforming path to overcome the critical aspects of traditional tourism models. The emerging discourse placed smart tourism amongst "the innovative projects to transform destination and re-envision tourism development models" (p. 10). Digital technologies, digital infrastructures, and human capital formation (including digital skills), also through the support of digital labs "to be prepared for a restart," were frequently discussed in relation to tourism redesign.

During the pandemic much attention was drawn to social media (Elkhwesky et al. 2022) as a form of consolidated technology on which hospitality businesses and DMOs significantly rely. With the travel bans, destinations were experienced online and social media communication was the only possible way to maintain an "online brand reputation" (Xiang and Gretzel 2010). Analyses of social media communication in the COVID-19 crisis, in the case of four Italian overtouristified cities, Rome, Milan, Florence, and Venice (Facebook and Instagram, Pasquinelli et al. 2021; Pasquinelli et al. 2022), shed light on the process of re-imagining these urban destinations, functionally and/or emotionally adapting the destination brand attributes and values to post-pandemic tourism markets. On the other side, sustainable destination propositions emerged, made possible by sudden and radical discontinuities, as a form of brand experimentation, whose long-term and structural effects need to be investigated.

1.5 Book Aim, Emerging Research Gaps, and Questions

This book aims to lay the foundation for sustainable-oriented innovation in smart tourism destinations, shedding light on the research gaps that have emerged in the literature and building novel research routes that are relevant to academics, practitioners, and policymakers. Although the research streams on technology-driven innovation and sustainable tourism have proliferated significantly in the last few decades, they remain fairly separated and self-contained. Conceptual overlaps must be tracked, and the emerging gaps and questions must be filled by combining the constructs and theories from the two research domains and conducting empirical explorations.

Certainly, in a time of deep economic and social crisis, as the post-pandemic time proved to be, technology investments can be at risk in times of crisis. Even though financed mainly by public authorities at regional, national, and supranational levels, correlated private stakeholders' investments regarding, for instance, human resources and relational capital needed to fully exploit the technological advancements might be less likely or even unrealistic in turbulent times.

There is an urgent need to frame new routes of academic investigation that combine theory and empirical evidence to address technological power towards sustainable-oriented innovation that is fruitful for policymakers. Sustainable tourism studies need further to boost the "so what?" agenda, reflecting on the policy and managerial implications of the wealth of theoretical and analytical contributions discussed in the academic debate. The analysis of technology deployment for pursuing sustainable development offers the opportunity to go in this direction by moving from the practice. Several policy initiatives, funding, and projects witness this real opportunity as tourism destinations are increasingly protagonists of significant public and private technological investments.

Several questions derive from the intersections of these gaps. First, how and to what extent is technology an effective enabling factor of innovation towards sustainable destination development? What are the co-determining conditions for boosting innovation processes that position the destination on more sustainable tracks of development? Second, adopting and using technologies in tourism destinations opens important questions about tourism stakeholders' engagement and their relationship with technology. How can technology boost tourists' perception and understanding of their contribution to tourism sustainability? How can technology help public and private decision makers manage and address tourism imbalances? How and to what extent can digital engagement support innovative processes? This book combines smart tourism and sustainable tourism studies in research efforts that frame and explain sustainable-oriented innovation dynamics in tourism destinations. The role of technology in the different forms of destination innovation and the factors co-determining innovations concerning product/service and experience, processes, managerial, and institutional dimensions, need to successfully integrate and fuel smart and sustainable tourism conceptualisation.

A central research question drives the book's investigation: how can DMOs and other key players leverage a smart tourism mindset to facilitate the combination of technology- and social-driven innovation towards sustainability? The book's theoretical and empirical analyses investigate technology- and social-driven innovation and draw attention to an emerging research stream concerning human–technology interaction which may enhance sustainability and societal wellbeing.

Technology's support for Destination Management Organisations (DMOs) in orchestrating the destination system towards sustainability opens up further research scenarios. An analysis of how DMOs can leverage ICT tools to pursue sustainable development, accessibility, inclusiveness, and wellbeing is part of this research agenda. Smart technologies provide an opportunity for knowledge sharing and information exchange (Jovicic 2019) that support evidence-based actions for addressing tourism balances and the mutual appreciation of the sustainability commitment amongst destination stakeholders. Developing this research avenue implies deepening knowledge of the modalities to support knowledge circulation and effective absorption by fostering and consolidating stakeholders' interpretative capacity. There is room to analyse the pivotal role and functions that the DMO can play (Sorokina et al. 2022), not only for tourism product innovation but also for process, managerial, and institutional innovations. A broader inquiry concerns technologies that boost destination governance transformations following technology-enhanced knowledge exchange and collaborative mindsets (Gretzel 2022).

Finally, studies on sustainability transitions need to consider lock-in contexts where different social, economic, cultural, and cognitive variables constrain the underpinning of sustainable development. How can technologies support the breaking of vicious cycles in tourism can development? What coexisting conditions allow destination stakeholders to trigger innovative mindsets, collaborative schemes, and products?

1.6 The Book's Structure

Following this chapter, stating the premises for the proposed research and framing its aim, the book is structured as follows. Chapter 2 frames innovation in the knowledge-based destination as a contextual and holistic result of the actors' co-evolutionary process, which exploits technological opportunities and levers social capital. It investigates innovation in smart tourism destinations from a broader perspective, combining technology- and social-driven innovation. A literature review provides a conceptual map of the academic investigation of technology-driven innovation drivers in the destination, considering: the actors involved in the innovation process (tourist, firm, local community, DMO, and institution), the ICT tools adopted, the role of social capital, and the types of innovation. Smart tourism (59%)—interpreted as a combination of smart destination, smart business ecosystem, and smart experience—represents the dominant form of innovation, followed by experience co-creation (28%). E-participative governance, social innovation, and creative tourism result as underinvestigated.

The chapter draws attention to the broader vision of smartness in tourism, which sheds light on the combination between technology, business, and society. It builds on the smart tourism mindset conceptualisation to reframe technology- and social-driven innovation. This picture of the academic investigation, in which technology enhances interaction and involvement, provides new areas of inquiry to frame and conduct theoretical research on human–technology interaction and wellbeing in destination co-creation, which can shape political agendas and destination management practices. The chapter calls for academic attention to innovative destination models combining smartness, cultural heritage, creativity, social innovation, and entrepreneurship, reshaping the role of each local actor, creating new business opportunities, and re-vitalising the cultural values of the local community.

Chapter 3 presents an analytical framework for smart and sustainable destination management. This framework builds on a critical review of the sustainable tourism literature outlining the advancements of the sustainable destination management research agenda. It relies on identifying and critically discussing conceptual integrations between the smart and sustainable destination theoretical debates. This framework proposes a broader interpretation of the ICT tools' adoption in the frame of sustainable tourism policies and actions. It integrates three layers of the academic investigation, which are not always observable in practice or a priori defined. The integration between intervention approaches (i.e. normative enforcement, visitor-oriented, and stakeholder-oriented destination management), strategic frameworks (i.e. a conservative transition and a radical framework), and visions (i.e. sustainable growth, sustainable development, and de-growth) represents a cutting-edge approach to smart and sustainable destination management. The proposed analytical framework opens new research avenues and can assist the design of sustainable policy evaluation frameworks, suggests how to monitor and measure the impacts of the actions undertaken, and may support the destination community towards the definition of sustainable tourism development goals.

Chapter 4 analyses a set of European destinations that have been investing in technology-driven innovation which drives tourist's immersive experience and enhances sustainable, inclusive, and accessible destinations. These destinations adopted smart technologies to enhance the quality of visitor experience and boost sustainable development, inclusiveness, and wellbeing. A widespread myopic approach to smart technologies deployment often limits the scope of technology adoption to consolidated and circumscribed uses in the practice. In contrast, the selected European destinations (including European Capitals of Smart Tourism) provide evidence of the capacity to combine smart technologies with the pursuit of the quality of the tourist's and the resident's place experience and to integrate technologies into broad local sustainable development strategies.

Genoa (Italy), the Faroe Islands, Dubrovnik (Croatia), Lyon (France), and Goteborg (Sweden) provide insights into different modalities to pursue more sustainable development through smart access and mobility, remote destination experiences, smart experience design, and knowledge ecosystem building. This chapter investigates the adopted tools and what functions and characteristics frame them, and it moves to discuss their adoption rationale and the underlying managerial approaches to the sustainable destination. Whilst normative enforcement does not significantly and exhaustively interpret the analysed cases, visitor-oriented destination management is at the core of several described projects. In addition, in some cases, stakeholder-oriented destination management complements the visitor-oriented approach to boosting an urban collective sustainability-oriented mindset and sustainability partnerships.

Chapter 5 draws preliminary conceptual advancements and sets the research agenda on sustainability-oriented innovation in smart tourism destinations. After the theoretical exploration and analysis of a set of European destinations in the previous chapters, this chapter states the key emerging traits of sustainability-oriented innovation in tourism destinations, opening up research routes and offering sustainability management and policy implications.

References

Accenture (2020) Waking up to a new reality: building a responsible future for immersive technologies.

Adams R, Jeanreanaud S, Bessant J, Denyer D, Overy P (2016) Sustainability-oriented innovation: a systematic review. Int J Manag Rev 18(2):180–205

Ali A, Frew AJ (2014) ICT and sustainable tourism development: an innovative perspective. J Hosp Tour Technol 5(1):2–16

Bach C, Scapin DL (2004) Obstacles and perspectives for evaluating mixed reality systems usability. In: Computer-aided design of user interfaces, pp 1–8

Bekele MK, Town C, Pierdicca R, Frontoni E, Malinverni ES (2018) A survey of augmented, virtual, and mixed reality for cultural heritage. J Comput Cult Heritage 11(2):7–36

Coenen L, Truffer B (2012) Places and spaces of sustainability transitions: geographical contributions to an emerging research and policy field. Eur Plan Stud 20(3):367–374

Collins-Kreiner N, Ram Y (2020) National tourism strategies during the COVID-19 pandemic. Ann Tour Res 89. https://doi.org/10.1016/j.annals.2020.103076

Colomb C, Novy J (2016) Protest and resistance in the tourist city. Routledge

Del Chiappa G, Pung JM, Atzeni M (2022) Factors influencing choice of accommodation during Covid-19: a mixed-methods study of Italian consumers. J Qual Assur Hosp Tour 23(4): 1037–1063

Dodds R, Butler RW (2019) Introduction. In: Dodds R, Butler RW (eds) Overtourism. Issues, realities and solutions. De Gruyter, Berlin

Dolnicar S, Zare S (2020) COVID19 and airbnb–Disrupting the disruptor. Ann Tour Res 83:102961

Edwards-Schachter M (2018) The nature and variety of innovation. Int J Innov Stud 2(2):65–79

Elkhwesky Z, El Manzani Y, Elbayoumi Salem I (2022) Driving hospitality and tourism to foster sustainable innovation: a systemic review of Covid-19-related studies and practical implications in the digital era. Tour Hosp Res 20:1–19

Fagerberg J (2018) Mobilizing innovation for sustainability transitions: a comment on transformative innovation policy. Res Policy 47:1568–1576

Fava N, Rubio SP (2017) From Barcelona: the pearl of the Mediterranean to bye bye Barcelona. In: Bellini N, Pasquinelli C (eds) Tourism in the city. Towards an integrative agenda on urban tourism. Springer, pp 285–296

Flavián C, Ibáñez-Sánchez S, Orús C (2019) The impact of virtual, augmented and mixed reality technologies on the customer experience. J Bus Res 100:547–560

Gossling S, Scott D, Hall M (2020) Pandemics, tourism and global change: a rapid assessment of COVID-19. J Sustain Tour 29(1):1–20

Gretzel U (2021) Conceptualizing the smart tourism mindset: fostering utopian thinking in smart tourism development. J Smart Tour 1(1):3–8

Gretzel U (2022) The smart DMO: a new step in the digital transformation of destination management organisations. Euro J Tour Res 30:302

Gretzel U, Fuchs M, Baggio R, Hoepken W, Law R, Neidhardt J, Pesonen J, Zanker M, Xiang Z (2020) E-tourism beyond COVID-19: a call for transformative research. Inf Technol Tour 22(2): 187–203

Hassan SB, Soliman M (2021) Covid-19 and repeat visitation: assessing the role of destination social responsibility, destination reputation, holidaymakers' trust and fear arousal. J Destin Mark Manag 19(March):100495

Higgins-Desbiolles F (2021) The 'war over tourism': challenges to sustainable tourism in the tourism academy after COVID-19. J Sustain Tour 29(4):551–569

Ioannides D, Gyimothy S (2020) The Covid-19 crisis as an opportunity for escaping the unsustainable global tourism path. Tour Geogr 22(3):624–632

Jovicic DZ (2019) From the traditional understanding of tourism destination to the smart tourism destination. Curr Issue Tour 22(3):276–282

Kahn KB (2018) Understanding innovation. Business Horizons 61(3):453–460

Karabulut G, Bilgin MH, Demir E, Doker AC (2020) How pandemics affect tourism: international evidence. Ann Tour Res 84:102991

Koens K, Postma A, Papp B (2018) Is overtourism overused? Understanding the impact of tourism in a city context. Sustainability (Switzerland) 10:1–15

Li J, Nguyen THH, Coca-Stefaniak JA (2021) Coronavirus impacts on post-pandemic planned travel behaviours. Ann Tour Res 86(January):1–5

Martín JMM, Martínez JMG, Fernández JAS (2018) An analysis of the factors behind the citizen's attitude of rejection towards tourism in a context of overtourism and economic dependence on this activity. Sustainability (Switzerland) 10(8)

Matiza T (2020) Post-COVID-19 crisis travel behaviour: towards mitigating the effects of perceived risk. J Tour Futures 8(1):99–108

Milano C (2017) Overtourism y Turismofobia. Tendencias globales y contextos locales. Ostelea School of Tourism & Hospitality, Barcelona

Milano C, Novelli M, Cheer JM (2019) Overtourism and tourismphobia: a journey through four decades of tourism development, planning and local concerns. Tour Plan Develop 16(4): 353–357

Minoia P (2017) Venice reshaped? Tourist gentrification and sense of place. In: Bellini N, Pasquinelli C (eds) Tourism in the city. Towards an integrative agenda on urban tourism. Springer, pp 261–274

Muler Gonzalez V, Coromina L, Galí N (2018) Overtourism: residents' perceptions of tourism impact as an indicator of resident social carrying capacity–case study of a Spanish heritage town. Tour Rev 73(3):277–296

OECD (2017) Science, technology and industry scoreboard 2017. The digital transformation. OECD Publishing, Paris, France

Oklevik O, Gössling S, Michael CH, Jens Kristian SJ, Ivar Petter G, McCabe S (2019) Overtourism, optimisation, and destination performance indicators: a case study of activities in fjord Norway. J Sustain Tour 27(12):1804–1824

Paramanathan S, Farrukh C, Phaal R (2004) Implementing industrial sustainability: the research issues in technology management. R D Manag 34(5):527–537

Pasquinelli C, Rovai S, Bellini N (2023a) Linking place brands and regional innovation: sustainable business strategies leveraging heritage. Reg Stud 1. https://doi.org/10.1080/00343404.2023.2187046

Pasquinelli C, Trunfio M (2020a) Overtouristified cities: an online news media narrative analysis. J Sustain Tour 28(11):1805–1824

Pasquinelli C, Trunfio M (2020b) Reframing urban overtourism through the smart-city lens. Cities 102:1–8

Pasquinelli C, Trunfio M (2020c) The European capitals of culture in a time of overtourism hazard. In: Cantoni L, De Ascaniis S, Elgin-Nijhuis K (eds) Proceedings of the Heritage, Tourism and Hospitality International Conference HTHIC2020. Università della Svizzera Italiana, Lugano, pp 159–170

Pasquinelli C, Trunfio M (2022) The missing link between overtourism and post-pandemic tourism. Framing twitter debate on the Italian tourism cities. J Place Manag Dev 15(3):229–247

Pasquinelli C, Trunfio M, Bellini N, Rossi S (2021) Sustainability in overtouristified cities? A social media insight into Italian branding responses to Covid-19 crisis. Sustain 13:1848

Pasquinelli C, Trunfio M, Bellini N, Rossi S (2022) Reimagining urban destination: adaptive and transformative city brand attributes and values in the pandemic crisis. Cities 124(May):103621

Pasquinelli C, Trunfio M, Punziano G, Del Chiappa G (2023b) Online tourism experiences: exploring digital and human dimensions in in-remote destination visits. J Hosp Mark Manag 32(3):385–409

Peeters PM, Gössling S, Klijs J, Milano C, Novelli M, Dijkmans CHS, ... Postma A (2018) Research for TRAN committee-Overtourism: impact and possible policy responses. European Parliament, Policy Department for Structural and Cohesion Policies

Pencarelli T (2020) The digital revolution in the travel and tourism industry. Inf Tech Tour 22(3): 455–476

Reinhold S, Laesser C, Beritelli P (2015) 2014 St. Gallen Consensus on destination management. J Destin Mark Manag 4(2):137–142

Séraphin H, Ivanov S, Dosquet F, Lajoinie B (2020) Archetypes of locals in destinations victim of overtourism. J Hosp Tour Manag 43:283–288

Séraphin H, Sheeran P, Pilato M (2018) Over-tourism and the fall of Venice as a destination. J Destin Mark Manag 9:374–376

Séraphin H, Zaman M, Olver S, Bourliataux-Lajoinie S, Dosquet F (2019) Destination branding and overtourism. J Hosp Tour Manag 38:1–4

Shafiee S, Ghatari AR, Hasanzadeh A, Jahanyan S (2019) Developing a model for sustainable smart tourism destination: a systemic review. Tour Manag Perspect 31:287–300

Sharma A, Nicolau JL (2020) An open market valuation of the effects of COVID-19 on the travel and tourismindustry. Ann Tour Res 83:102990

Sigala M (2020) Tourism and COVID-19: impacts and implications for advancing and resetting industry and research. J Bus Res 117:312–321

Sorokina E, Wang Y, Gyall A, Lugosi P, Torres E, Jung T (2022) Constructing a smart destination framework: a destination marketing organisation perspective. J Destin Mark Manag 23 (March):100688

Stankov U, Gretzel U (2020) Tourism 4.0 technologies and tourist experiences: a human-centered design perspective. Inf Tech Tour 22(3):477–488

Talwar S, Kaur P, Nunkoo R, Dhir A (2022a) Digitalisation and sustainability: virtual reality tourism in a post pandemic world. J Sustain Tour 1. https://doi.org/10.1080/09669582.2022.2029870

Talwar S, Srivastava S, Sakashita M, Islam N, Dhir A (2022b) Personality and travel intentions during and after the Covid-19 pandemic: an artificial neural network (ANN) approach. J Bus Res 142:400–411

Trunfio M, Campana S (2020) Innovation in knowledge-based destination: technology-driven versus social-driven. Int J Knowl Based Dev 11(2):76–199

Trunfio M, Campana S (2019) Drivers and emerging innovations in knowledge-based destinations: towards a research agenda. J Destin Mark Manag 14:100370

Trunfio M, Jung T, Campana S (2022) Mixed reality experiences in museums: exploring the impact of functional elements of the devices on visitors' immersive experiences and post-experience behaviours. Inf Manag 59:1–16

Trunfio M, Pasquinelli C (2021) Smart technologies in the Covid-19 crisis: managing touris flows and shaping visitors' behaviour. Euro J Tour Res 29:2910

UNWTO (2020) Covid-19 and tourism. 2020: a year in review. UNWTO

van der Have RP, Rubalcaba L (2016) Social innovation research: an emerging area of innovation studies? Res Policy 45:1923–1935

Wong IA, Lu MV, Lin S, Lin Z (2022) The transformative virtual experience paradigm: the case of Airbnb's online experience. Int J Contemp Hosp Manag. https://doi.org/10.1108/IJCHM-12-2021-1554

Xiang Z, Gretzel U (2010) Role of social Media in Online Travel Information Search. Tour Manag 31:179–188

Zhang H, Song H, Wen L, Liu C (2020) Forecasting tourism recovery amid COVID-19. Anna Tour Res 87:1–16

Zhang J, and Qiu H (2022) Window to the destination: tourists' local experience via "online experiences" on Airbnb amid the pandemic. Paper presented at the ENTER 2022 eTourism Conference

Zheng D, Luo Q, Ritchie BW (2021) Afraid to travel after COVID-19? Self-protection, coping and resilience against pandemic 'travel fear. Tour Manag 83:104261

Zhu J, Cheng M (2021) The rise of a new form of virtual tour: airbnb peer-to-peer online experience. Curr Issue Tour:1–6

Chapter 2
Blending Technology-Driven and Social-Driven Innovation in Smart Tourism Destinations

2.1 Innovation: Key Streams of Conceptualisation

Defining innovation as a broader and encompassing field has represented a stimulating but risky challenge for academics, supranational institutions, policymakers, and consultants. Innovation has been proposed as an overarching solution for global economic growth and sustainability, driving political agendas (Fagerberg 2018). Although an increasing number of academic papers, reports, and White Papers demonstrate growing attention, "innovation" remains an umbrella term and a buzzword involving various concepts, definitions, perspectives, and controversial application areas.

The term innovation has been over-simplified and over-used in association with several terms, such as technological, digital, open, disruptive, transformative, creative, R&D-driven (internal), user-driven (external), public, inclusive, accessible, sustainable, and green. Innovation emerged in the early 1960s as a scientific discipline and has been developing rapidly in the last few decades, with the significant influence of historical and socio-cultural contexts (Fagerberg and Verspagen 2009).

The prevailing definitions followed Schumpeter's seminal paper (1934), which interpreted innovation as an entrepreneurial process generating something new (products, production processes, markets, resources, and organisation) or a combination of new and existing knowledge. Schumpeter remained the most influential scholar in the academic debate for several decades.

One of the most widespread definitions adopted in international and institutional contexts has been proposed in the Oslo Manual (OECD/Eurostat 2005). Following Schumpeter's seminal paper (1934), it defined innovation as a new or significantly improved product (good or service), process, a new marketing method, or a new organisational method in business practices, workplace organisation, or external relations. Several innovations spread throughout organisations and networks, leveraging creativity, knowledge, competencies, and skills.

C. Pasquinelli, M. Trunfio, *Sustainability-oriented Innovation in Smart Tourism*, Tourism on the Verge, https://doi.org/10.1007/978-3-031-33677-5_2

Investigating the nature and variety of innovation, Edwards-Schachter (2018) identified salient types of innovation relevant for researchers and practitioners, combining technological and non-technological innovations. She considered consolidated kinds of innovation (product, process, and service), radical and disruptive innovation, business model innovation, new and recombined forms of technological innovation as the result of the fourth industrial revolution, and design-driven innovation (Edwards-Schachter 2018).

The consolidated innovation paradigm focused mostly on science, technology, and the economy, giving limited responses to complex processes and changes in the global society and economy. Recently, research has devoted attention to social and responsible innovation as challenging transformations that achieve sustainable development (Edwards-Schachter 2018). Social innovation can be defined as novel social technologies that create new social value, change social relationships, and solve socially relevant problems (van der Have and Rubalcaba 2016). Social innovation's effects turn into "social technology" forms that can generate positive spillover or incubators in local contexts (Neumeier 2017; When and Evers 2015).

The relationship between sustainability and innovation has been widely investigated (Adams et al. 2016). Sustainable development policies and practices can leverage innovation as a challenge to provide good models (Fagerberg 2017). Frugal innovation is gaining increasing attention in the academic debate as a form of sustainable innovation. It covers consolidated forms of innovation (i.e., product, service, process, or business model) by exploiting resource-scarce solutions (financial, technological, material, or other resource solutions) (Hossain 2018; Rosca et al. 2017). Social, responsible, and frugal innovation embraces the notion of changing mindsets (Kahn 2018) and solving societal needs and imbalances, creating hybrid innovation as "mixed modes of innovation" (OECD 2017, p. 154) which recombine technology, society, and economy.

2.2 Innovation in Tourism and Hospitality Research

As a multidimensional phenomenon, innovation acquires a different connotation in the different disciplines, including diverse concepts, definitions, and perspectives, some of which co-exist in emergent or consolidated fields of innovation studies (Edwards-Schachter and Wallace 2017). Innovation represents a hot topic in tourism and hospitality research, challenging academics to carry out a multidisciplinary investigation and influencing the political agenda around the world. Several theoretical and empirical studies have analysed diverse aspects of the controversial and multifaceted innovation phenomenon. They have proposed conceptual frameworks and interpretative models, adopting specific angles (e.g. technologies) and involving various actors (destination management organisations, policymakers etc.).

In continuity with the consolidated paradigm of innovation, research on innovation in the tourism and hospitality domains has increased significantly in the last decade, following Hjalager's (2010, 2015) papers. Hjalager's categorisation integrated established forms of innovation *outside* the tourism domain with innovation *specific to* that domain (Hjalager 2010): product or service innovations; process innovations; managerial innovation; and institutional innovation. Hjalager (2015) identified 100 innovations that have transformed tourism. These include: increasing the social and physical effects of the processes that create tourist products and services (and thus experiences); increasing the productivity and efficacy of tourism enterprises; building new destinations; improving mobility to and within destinations; increasing opportunities to transfer and share information; changing forms of institutional logic and power relations.

Several systematic literature reviews have been proposed recently, investigating different but complementary angles of innovation in tourism and hospitality. Reviewing the heterogeneous literature on innovation in tourism and hospitality, Gomezelj's study (2016) analysed 152 articles. It defined a coherent and integrated theoretical framework in which nine co-citation networks or clusters of papers are: fundamental studies (literature reviews), Resource-Based View and competitive advantage, organisational studies, networking, innovation in service, innovation systems, knowledge, management of organisational innovation, and technology. These research fields have been discussed at three levels of analysis: the micro or firm level; the macro level (society, regions, and tourism destinations); and the general level (innovation systems, often involving collaboration between diverse institutions).

As tourism is a complex field of research, academics have called for interdisciplinary research (Pikkemaat et al. (2018) and have proposed a systematic literature to identify emerging fields of tourism innovation research. Several areas of investigation have been detected: public policy, experience, technology, knowledge, socio-environmental, network-cooperative, and organisation. The study proposed four pillars of tourism innovation research: processes, context configuration (e.g. organisation, destination, and network), knowledge and technology (diffusion and absorption), and eco-innovations (Pikkemaat et al. 2018). They conclude that European agendas, policymakers (also levering on public funding), and destination management organisations drive and facilitate innovation involving enterprises and stakeholders.

Since 2018, academic interest in tourism innovation has increased rapidly, and the annual production of research on innovation has more than doubled. Tourism and hospitality research received new stimuli and contaminations from other disciplines. A co-word analysis of a recent systematic literature review (Işık et al. 2022) defined a network of themes driving research on tourism innovation. The prevalent and most significant advances in tourism innovation research are: ICTs, including technology acceptance and the sharing economy; knowledge management, including absorptive capacity, knowledge sharing, and knowledge transfer; social capital; networks and collaboration; entrepreneurship, including firm performance; sustainable development, including sustainable tourism, nature-based tourism and rural tourism; and cultural tourism (Işık et al. 2022).

2.3 Innovation in Knowledge-Based Destination

Innovation is a controversial and often ambiguous term, and when we try to investigate it in the field of the destination, the complexity increases and requires diverse angles of inquiry and a multidisciplinary approach. The prevailing literature considers diverse forms of tourism innovation—which mainly lever on technologies and marginally on social processes—as primary drivers of new destination models that involve various stakeholders and enhance sustainability (Kuščer et al. 2017; Narduzzo and Volo 2018; Ozseker 2018; Paget et al. 2010; Pike and Page 2014; Pikkemaat et al. 2018; Trunfio and Campana 2019).

Academic investigation has emphasised the disruptive role of information and communication technologies (ICTs) and the power of smartness in tourism innovation. Technology has been presented as contributing to sustainable destination development, driving stakeholder engagement and interaction between tourists and destinations, which allows valuable experience co-creation (Buhalis and Law 2008; Femenia-Serra et al. 2019; Gretzel et al. 2015; Shafiee et al. 2019; Sigala 2018; Trunfio and Campana 2019).

Destination management organisations (DMOs) and policymakers invest in ICT tools and digital innovation as a panacea for tourism imbalances and challenges for the destination. They interpret technological innovation as the driver of change that connects enterprises, local communities, and tourists, facilitating smart destination projects, including innovative technology for sustainability (Ali and Frew 2014a, b; Buhalis and Amaranggana 2015).

The literature debates the consolidated tourism innovation paradigm, calling for a critical approach to overcome disciplinary myopia and business bias that characterises technology research (Sigala 2018) and draws future research scenarios, considering the cross-fertilisation with other theoretical domains (Trunfio and Campana 2019). New ways of thinking about the relationship between emerging technologies and societal change—including the investigation of value co-creation and co-destruction involving various actors—can advance a critical theoretical approach to ICTs in tourism (Paget et al. 2010; Koo et al. 2017; Sigala 2018; Stankov and Gretzel 2020; Tribe 2007; Trunfio and Campana 2019; Trunfio et al. 2022).

Academics are called on to reframe innovation in the evolutionary process of tourism destinations, overcoming the myopic vision of the disruptive technological power in tourism and embracing the complexity of human–technology interaction, which requires a human-centred approach in the destination investigation (Stankov and Gretzel 2020; Trunfio and Campana 2019; Trunfio et al. 2022). Diverse and interactive social conditions, cultural values, and actors' behaviours determine the complexity of the destination as a complex social system (Beritelli and Bieger 2014), posing several pitfalls and challenges in framing innovation processes in the tourism domain.

The evolutionary process of innovation is contextual and embedded in geographical space; it involves several actors in the dynamic spiral of knowledge sharing, collective innovativeness, and spillover effects (Trunfio and Campana 2019). Overcoming the traditional paradigm of innovation in tourism, Trunfio and Campana (2019) interpreted innovation as embedded in the complexity of a tourism destination and as a knowledge-driven phenomenon.

They adopted Nonaka and Konno's "Ba" concept (1998), which designs physical, virtual, and cognitive spaces to create, develop, codify, share, and disseminate collective knowledge, which drives innovation in the destination (Trunfio and Campana 2019). Consequently, the knowledge-based destination is nurtured in specific local contexts where several actors interact. Local contexts represent a repository of spatial and cross-sectorial knowledge generation and dissemination, which create conditions for collective innovativeness which determine pervasive and holistic innovation (Boekema et al. 2000; Carvalho and van Winden 2017; Flagestad et al. 2005; Roper and Love 2018; Trunfio and Campana 2019; Yigitcanlar and Lönnqvist 2013).

In the broader theoretical framework of knowledge management, absorptive capacity represents the antecedent of innovation that influences innovativeness in local contexts; investigating absorptive and organisational learning activities can help detect key aspects that transform internal and external knowledge in innovation (Schuhbert, 2021). The processes of knowledge sharing and dissemination in the destination embrace several aspects such as (Hoarau and Kline 2014; Ozseker 2018; Racherla et al. 2008; Sheehan et al. 2016): developing and sharing knowledge; integrating tacit and explicit knowledge; socialising and transfer knowledge; transforming knowledge in innovation; adopting specific policies and strategies to enhance and disseminate innovation.

Destination can be seen as a complex knowledge-based social system that co-creates (or co-destructs) value by generating a co-evolutionary process of innovation (Van Assche et al. 2013; Trunfio and Campana 2019; Tuli et al. 2019). Destination innovation, as a knowledge-driven phenomenon embedded in specific local contexts, is driven by actors (DMO, institutions and political actors, local firms, local community, and tourists) who are connected through social and technological platforms (Trunfio and Campana 2019). They enhance actors' interaction and collective knowledge creating pervasive and holistic innovations, such as smart tourism, experience co-creation, e-participative governance, social innovation, and creative tourism (Trunfio and Campana 2019) (Fig. 2.1).

Innovation in knowledge-based destination, as a contextual and holistic result of the co-evolutionary process of several actors, exploits the technological opportunities and is nurtured by social capital. Two broader forms of innovation can be framed: technology-driven and social-driven (Trunfio and Campana 2020).

Technology-driven innovation considers ICTs as fundamental infrastructures allowing actors' collaboration and integration; facilitating knowledge sharing and dissemination; creating forms of smart sources of information for visitors and smart experiences; connecting a destination with distant stakeholders; introducing a new business model; facilitating e-participative strategic approaches in destination management, combining offline and online stakeholder engagement. Technology-driven innovation can be unique and involve a few destination actors (e.g. a start-up

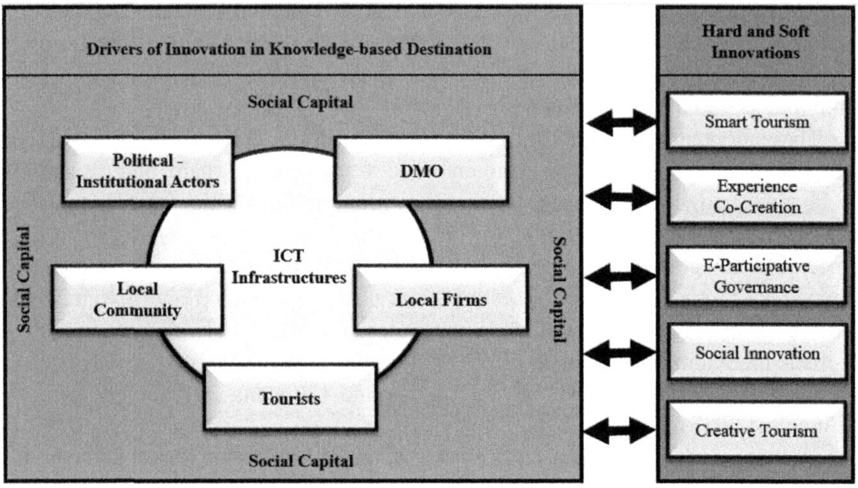

Fig. 2.1 Drivers and innovations in knowledge-based destination. Source: Adapted by Trunfio and Campana (2019, p. 5)

introducing an innovative app for virtual reality in the destination) to involve several actors, becoming inclusive and activating the purpose of the smart destination.

Although technology-driven innovation has been presented as the "unique and ready" solution for innovation processes, it offers several disadvantages. It requires technology acceptance, adoption, and intention to use, which limits the effectiveness of the application and generates conflicts in human–technological interaction, such as information overload, technostress, and difficulties of use (Stankov and Filimonau 2019; Trunfio et al. 2022).

Socially-driven innovation levers on social capital and involves the local community and businesses in bottom-up innovation processes. Social capital nurtures social-driven innovation and creates structural, cognitive, and relational conditions to include all destination actors in the innovation process. It facilitates the strategic integration of common interests and goal alignment, encourages personal relationships and trust, and enables absorptive capacity as pre-conditions for inclusive and bottom-up innovation.

Social innovation is a novel "social technology" that creates new social value, changes social relationships, and solves socially relevant problems and imbalances (van der Have and Rubalcaba 2016). It reshapes the social capital structure of the destination after several years. It promotes multi-level social changes and inclusive destination models; it allows us to solve social imbalances (poverty, criminality, unemployment, marginalisation etc.) and stimulates social awareness of community issues and proactive behaviour, increasing collective societal welfare (Trunfio and Campana 2020).

2.4 Social-Driven Innovation

Social-driven innovation represents a co-evolutionary process involving the social structure of the destination; bottom-up actors' participation and social capital play a dominant role in destination changes and innovation (Trunfio and Campana 2020). Diverse destination actors can facilitate (or limit) innovation by leveraging bottom-up or top-down processes and enhancing stakeholder engagement. Virtuous circles of knowledge (generation, sharing, and dissemination), collective innovativeness, and pervasive innovation (Boekema et al. 2000; Camisón and Monfort-Mir 2012; Trunfio and Campana 2019) generate opportunities to reframe and reshape destination identity and destination value proposition for the diverse stakeholders activating social changes.

Although the literature investigated the role of destination actors extensively, covering several aspects of destination management (including network analysis), the power of social capital in nurturing innovation and driving the destination evolutionary processes remains under-investigated. Empirical research explores the role of social capital in tourism research as a platform nurturing innovation and facilitating sustainable destination development, cultural regeneration, and social inclusion (Go et al. 2013; Trunfio and Campana 2020). Diverse patterns of destination innovation can emerge considering several aspects during the four phases of the co-evolutionary process, such as: who is the key player and which roles are assumed in driving innovation processes; which is the power and the legitimation of public and/or private actors; how and how much are stakeholders involved in the process; which form of networking and collaboration can facilitate innovation etc.

2.4.1 *The Role of Destination Actors*

The process of innovation in the knowledge-based destination is socially embedded. Several aspects influence the destination co-evolutionary processes and their spillover effects (Asheim et al. 2007; Baggio and Cooper 2010; Carrillo 2014; Liu and Lin 2012; Van Assche et al. 2013). They are the number and variety of destination actors, the density and the intensity of social relationships, knowledge sharing and dissemination, capabilities, competences, and skills.

Interactions between actors (DMOs, institutions, and political actors, local firms and the local community, tourists)—as well as networking, coordination, and collaboration—boost the complexity of knowledge-driven innovation processes. Actors' interactions co-create and transform knowledge, resources, and experiences into several forms of innovation (Racherla et al. 2008; Sheehan et al. 2016; Trunfio and Campana 2019).

Considering the role of key actors, social-driven innovation can be described as a co-evolutionary process based on four relevant phases (Trunfio and Campana 2020):

social problematisation, expression of common interests, coordination actions, and success or failure. The innovation process in the destination has as its first phase social problematisation, during which key actors (the local community, NGOs, firms, DMOs, policymakers etc.) become aware of social imbalances, destination constraints, challenges, and opportunities etc. They express the will to seek a solution and activate forms of collaboration to realise specific projects (cultural regeneration, human capital development etc.). Social awareness and contributing to community issues enhance proactive community behaviours and collective societal welfare (Nespolo et al. 2018; Sanzo-Perez et al. 2015).

Consolidation of the first phase opens the second phase, called the expression of common interest. Awareness of common interests allows the adoption of a shared vision. It creates conditions for a social factory that participates in local innovation projects. It facilitates the strategic integration of common interests and goal alignment (Van Assche et al. 2013).

The third phase is centred on coordination and social learning, and engagement. In this phase, new practices of knowledge sharing and dissemination based on personal relationships and trust are developed (Martínez-Pérez et al. 2019). They allow us to attract creative and intellectual talents (Mengi et al. 2017; Pique et al. 2019) and transform the destination into a creative learning and experience system (Carvalho and van Winden 2017).

Finally, success (or failure) can intervene after a long period. In terms of success, social-driven innovation always turns into co-created social value, such as generating positive spillover effects or incubator actions within the local context, or which can reshape the social structure and identity of the destination, promoting new destination models or developing entrepreneurship.

One of the leading research challenges is exploring how diverse actors create power, collaboration, and networking, facilitating innovation and sustainable development in the destination. Unusual stakeholders can drive innovation in the destination, as in the case of the parish priest Antonio Loffredo. He has guided the co-evolutionary process of innovation in the "Rione Sanita" of Naples (Italy) to overcome social imbalances (local poverty, unemployment, and crime), levering on bottom-up processes which involve the local community (Trunfio and Campana 2020). He interpreted the problematisation phase and drove innovation in the destination, leveraging on leadership to facilitate social reshaping (Saito and Ruhanen 2017; Zehrer et al. 2014) and stakeholder engagement, stimulating the assumption of specific forms of collective power amongst destination actors (Beritelli et al. 2016; Saito and Ruhanen 2017). Projects of urban-heritage regeneration and human valorisation created a social fabric of entrepreneurship and innovation, overcoming the absence of institutions, political actors, and DMOs.

2.4.2 The Role of Social Capital

Social capital has received incremental attention as a multidisciplinary topic (Adler and Kwon 2002). Norms, values, beliefs, tacit knowledge, trust, and alternative forms of interaction determine social capital (Coleman 1988; Inkpen and Tsang 2005; Nahapiet and Ghoshal 1998; Putnam 1993).

Tourism research has investigated social capital as a driver of innovation (Go et al. 2013; Lee 2015; Macbeth et al. 2018; Trunfio and Campana 2019). In the knowledge-based destination, social capital configures the DNA, which enables (or limits) several aspects of the evolutionary path of innovation such as: local community awareness, a common vision, goal alignment, and shared strategies; learning processes, competencies, and skills development (including for young people); leadership, collaboration, networking etc.; and entrepreneurship incubator and development etc. Tourism management literature configured the social capital of destination through three dimensions, namely (Go et al. 2013; Trunfio and Campana 2020): structural, cognitive, and relational.

The structural dimension considers social capital, adopting the destination network perspective. Investigating networks (key actors' role, configuration, and stability) allows us to explore processes of generating, transforming, and sharing knowledge and their impact on boosting collective innovation (Inkpen and Tsang 2005; Liu and Lin 2012; Martínez-Pérez et al. 2019; Trunfio and Campana 2019).

Identifying primary, secondary, and salient actors' roles in networking allows us to design power and legitimation (Baggio and Cooper 2010; Beritelli et al. 2016). Key actors' power and legitimation determine the primary role and serve as an impulse for innovation. The secondary role emerges as the final effects or results of the entire innovation process. Finally, a salient role integrates the primary and secondary roles and assumes the leadership of social-driven innovation. Other actors can be related by hierarchical or non-hierarchical authority.

The cognitive dimension influences the process of knowledge sharing and dissemination amongst the actors of the destination and networks (Zehrer et al. 2014). Several aspects shape the cognitive dimension and drive innovation, such as: shared culture, codes, and languages; a sense of belonging to the place and a common vision; leadership and mutual benefits; destination strategy definition and implementation. Summarising, the cognitive dimension of social capital creates local embedded conditions to develop and share knowledge that drives the co-evolutionary innovation process in the destination.

Norms, obligations, and trust shape the relational dimension of social capital, aligning behaviours to facilitate knowledge sharing and nurture innovation. Norms and obligations define mutual forms of control, sanction, or incentive for network actors. Trust affects network attitudes towards common goal alignment, reducing destination social complexity, and improving interaction and collaboration.

Differences in social capital, as well as the diverse attitude of key actors to innovativeness (and their power and legitimation) enhance (or limits) the processes of knowledge generation and sharing and drive innovation in local contexts (Ozseker 2018; Roper and Love 2018; Trunfio and Campana 2019). They reinforce stakeholder engagement practices in the destination and generate forms of pervasive innovation in the destination (Trunfio and Campana 2019).

2.5 Technology-Driven Innovation

Technology-driven innovation considers ICTs as fundamental infrastructures to reinforce the co-evolutionary innovation process in the destination. It guides a shift from the monologue of the actors' destination to multiple interconnections between supply and demand, offering innovative tourism experiences and overcoming environmental and social issues.

The literature attributed to ICT the role of essential infrastructure to nurture pervasive innovation processes in destination, supporting new forms of knowledge and actors' collaboration and generating new products (Buhalis 2019; Buhalis and Law 2008; Buhalis and O'Connor 2005; Gretzel et al. 2000; Neuhofer et al. 2012; Racherla et al. 2008; Stamboulis and Skayannis 2003; Trunfio and Campana 2019). Opportunities and challenges of destination innovation have been proposed, considering actors' engagement, interaction practices of knowledge sharing, and dissemination through hard and/or soft technological interdependences (Marasco et al. 2018; Trunfio and Campana 2019; Trunfio and Della Lucia 2019).

Buhalis's (1998) seminal work recognised tourism as an information-intensive industry, forecasting information technology (ICT) as a critical innovation driver to manage the future challenges that would arise. Indeed, Web-based technologies have played a crucial role in the paradigm shift, which has led to organisations, destinations, and tourists communicating, collaborating, managing, and experiencing tourism products together with new interactive interfaces and without barriers to social communication (Buhalis and O'Connor 2005; Munar 2011, 2012; Neuhofer et al. 2012; Stamboulis and Skayannis 2003).

Since Buhalis's (1998) work, several novel studies have intensified attention towards ICTs and have amplified its concept with the expression of disruptive technologies (Buhalis 2019; Buhalis et al. 2019), identifying or testing specific applications with empirical analysis methods (Aarstad et al. 2015; Buhalis et al. 2019; Buhalis and O'Connor 2005; Buhalis and Sinarta 2019; Corfu and Kastenholz 2005; Gretzel et al. 2000; Marasco et al. 2018; Racherla et al. 2008; Stamboulis and Skayannis 2003). The literature has explored how ICT has been tailored to individual or organisational propensities and environmental factors to gain technology acceptance and to manage innovation in such dynamic and unpredictable environments as knowledge-based destinations (Aarstad et al. 2015; Baggio and Cooper 2010; Lee 2015; Roper and Love 2018; Russo-Spena et al. 2017).

The literature has investigated how ICTs manifest their disruptive role in knowledge generation, sharing, and dissemination (Centobelli and Ndou 2019; Racherla et al. 2008; Sheehan et al. 2016; Sigala 2018; Stankov and Filimonau 2019) and in facilitating networking, coordination, and collaboration, improving the efficiency of tourism resource management for destination intelligence (Buhalis and Amaranggana 2013; Centobelli and Ndou 2019; Del Chiappa and Baggio 2015; Femenia-Serra et al. 2019; Gretzel et al. 2015; Ivars-Baidal et al. 2019; Del Vecchio et al. 2018; Del Vecchio and Passiante 2017). They may assist social capital building by enhancing learning capacity, anticipation, and self-organisation of the destination (Boes et al. 2015; Buhalis and Amaranggana 2013; Del Chiappa and Baggio 2015; Ivars-Baidal et al. 2019; Del Vecchio et al. 2018; Del Vecchio and Passiante 2017), transforming the traditional destination into a digital and/or smart innovation ecosystem.

Forms of technology-driven innovations in destination enhance: new destination models based on the digital and/or smart ecosystem (Buhalis 2019; Buhalis and Amaranggana 2015; Del Chiappa and Baggio 2015; Femenia-Serra et al. 2019; Gretzel et al. 2015); personalisation processes and experience co-creation (Buhalis et al. 2019; Buonincontri and Micera 2016; Neuhofer et al. 2012; Stamboulis and Skayannis 2003); participative approaches of governance in destination management to integrate offline and online stakeholder engagement (Barandiarán et al. 2019; Cabiddu et al. 2013. Sigala and Marinidis 2012; Trunfio and Della Lucia 2019); and processes of urban and cultural revitalisation and regeneration promoted by forms of social entrepreneurship (Della Lucia and Trunfio 2018; Rastrollo-Horrillo and Rivero Díaz 2019) or creative behaviours of the local community (Lindroth et al. 2007); and immersive experiences in cultural heritage and other tourism fields, levering on virtual, augmented, and extended reality (Trunfio et al. 2022; Yung and Khoo-Lattimore 2019).

2.6 Mapping Technology-Driven Innovation in Tourism Research: A Literature Review

Although the disruptive power of technology-driven innovation in tourism destinations has been widely recognised, the academic investigation remains fragmented, requiring further research to map and integrate diverse aspects of this multifaced phenomenon (Gomezelj 2016; Hjalager 2010; Rodriguez et al. 2014; Trunfio and Campana 2019).

A systematic literature review approach can advance knowledge in the field of technology-driven innovation, offering insights and guidance for future research and a dashboard for destination management practices and political agendas. It systematically detects destination innovation in the academic debate, investigating how scholars frame forms and drivers of technology-driven innovation in tourism destinations. The research is built on the rigorous, transparent, and reproducible protocol of the systematic quantitative literature review as a valuable method for emerging research streams (Pickering and Byrne 2014; Torraco 2016; Xin et al. 2013; Yung and Khoo-Lattimore 2019).

The research adopts the interpretative lens of Trunfio and Campana's (2019) theoretical framework, which considers technology innovation in the destination as a knowledge-driven phenomenon nurtured in specific local contexts and driven by destination actors. It identifies the main drivers of the forms of pervasive innovations (smart tourism, experience co-creation, e-participative governance, social innovation, and added creative tourism), in which social and technological dimensions have been combined.

Five typologies of actors (institutions, destination management organisations (DMOs), firms, local communities, and tourists) and two platforms of support (ICTs and social capital) have been considered to influence innovation in the destination. Introducing tourists (as a destination innovation driver) and creative tourism (as an innovation form) in the model allows us to remove the theoretical limitation considered in Trunfio and Campana's (2019) previous work, clarifying the pervasive and holistic innovation phenomenon in a tourism destination.

The systematic literature review protocol has been conducted to identify 213 papers in Elsevier's Scopus. It is considered the most extensive academic database in comparison to the EBSCO host (Hospitality and Tourism Complete), Science Direct (Elsevier), Emerald, Web of Science, ProQuest, and Sage, according to its advanced search capabilities (Yang et al. 2017). It used four search strings—"-destination", "tourism", "innovat*", and "technolog*" —searching under title, abstract, and keywords. The research selected papers in which innovation is technology-driven (word "technolog*"), considering the ICT tools promoting innovation in the destination.

The 120 articles were screened considering English-language peer-reviewed journal articles, to safeguard the quality and effectiveness of the review. We searched by article and document types for papers contributing to defining tourism destination innovation. Selected papers in which ICT tools have been investigated can present social capital as a complementary platform for innovation or not.

A preliminary analysis excluded 33 of 120 articles because their subject matter was not closely related to the topic of destination management. They cover these topics: entrepreneurial management practices, tour operators and tourism travel agencies, global tourism trends, management practices of water and subterranean resources, European activities planning, and management practices of navigation and communication.

The systematic literature review has allowed us to highlight the significance of the topic at hand. Table 2.1 summarises the 87 articles on "destination", "tourism", "innovat*", and "technolog*" which were included, which dealt with critical themes in tourism destination innovation. It shows the author/s, year, title, and journal for each article.

Analysis of the authors and their titles in Table 2.1 allows us to identify four works that have registered over 200 author's citations that have obtained the qualification of seminal articles in accordance with the values of timeliness, utility, and quality expressed by the scientific community (Okubo 1997; Walman 2016). These were: first, Hjalager's (2010) paper, "A review of innovation research in tourism", with 516 author citations; second, Buhalis's (1998) paper, "Strategic use

Table 2.1 The 87 articles on "destination", "tourism", "innovat*", and "technolog*"

Authors	Year	Title	Source title
Barandiarán, Restrepo & Luna	2019	Collaborative governance in tourism: Lessons from Etorkizuna Eraikiz in the Basque Country, Spain	Tourism Review
Buhalis	2019	Technology in tourism-from information communication technologies to eTourism and smart tourism towards ambient intelligence tourism: A perspective article	Tourism Review
Buhalis, Harwood, Bogicevic, Viglia, Beldona & Hofacker	2019	Technological disruptions in services: Lessons from tourism and hospitality	Journal of Service Management
Bulti, Ray & Bhuyan	2019	Smart tourism system architecture design using the Internet of Everything (IOE) over cloud platform	International Journal of Innovative Technology and Exploring Engineering
Chung, Tyan & Lee	2019	Eco-innovative museums and visitors' perceptions of corporate social responsibility	Sustainability
Garcia-Sanchez, Siles & Vazquez-Méndez	2019	Competitiveness and innovation: Effects on prosperity	Anatolia
Gomez-Oliva, Alvarado-Uribe, Parra-Merono & Jara	2019	Transforming communication channels to the co-creation and diffusion of intangibles heritage in smart tourism destination: Creation and testing in Ceutì (Spain)	Sustainability
Lam-Gonzalez, Leon & de Leon	2019	Assessing the effects of the climatic satisfaction on nautical tourists' on-site activities and expenditure decision	Journal of Destination Marketing and Management
Liu & Nijkamp	2019	Inbound tourism as a driving force for regional innovation: A spatial impact study on China	Journal of Travel Research
Maestro & Dumlao	2019	Romblon Islands into a smart tourism destination through point of interest recommender, augmented reality, and near field communication: A proposal	International Journal of Innovative Technology and Exploring Engineering
Muliawaty, Alamsyah & Loupias	2019	The role of new public service model in developing tourist destination in Bandung city: A new paradigm for domestic tourism industry	International Journal of Scientific and Technology Research

(continued)

Table 2.1 (continued)

Authors	Year	Title	Source title
Nilsson	2019	Urban bicycle tourism: Path dependencies and innovation in Greater Copenhagen	Journal of Sustainable Tourism
Rostrollo-Horrillo & Rivero Diaz	2019	Destination social capital and innovation in SMEs tourism firms: An empirical analysis in an adverse socio-economic context	Journal of Sustainable Tourism
Sanz-Ibanez, Lozano & Anton Clavé	2019	Brokers in a destination's knowledge networks	Journal of Destination Marketing and Management
Weaver & Moyle	2019	"Tourist stupidity" as a basic characteristic of "smart tourism": Challenges for destination planning and management	Tourism Recreation Research
Xu, Nash & Whitmarsh	2019	Big data or small data? A methodological review of sustainable tourism	Journal of Sustainable Tourism
Tshipala, Coetzee & Potgieter	2019	Sustainable indicators for adventure tourism destination: A case of Waterval Boven	African Journal of Science, Technology, Innovation and Development
Çakar, Kalbaska, Inanir & Şahin Ören	2018	eVisa's impacts on travel and tourism: The case of Turkey	Journal of Hospitality and Tourism Technology
da Costa Liberato, Alén-González & de Azevedo Liberato	2018a	Digital technology in a smart tourist destination: The case of Porto	Journal of Urban Technology
da Costa Liberato, Alén-González & de Azevedo Liberato	2018b	Smart tourism destination triggers consumer experience: The case of Porto	European Journal of Management and Business Economics
Ghaderi, Hatamifar & Henderson	2018	Destination selection by smart tourists: The case of Isfahan, Iran	Asia Pacific Journal of Tourism Research
Lodeiro-Santiago, Santos-González, Caballero-Gil, Caballero-Gil & Herrera-Priano	2018	Novel guidance CPS based on the Fat-Beacon protocol	Applied Sciences (Switzerland)
Mofokeng & Matima	2018	Future tourism trends: Virtual reality-based tourism utilising distributed ledger technologies	African Journal of Hospitality, Tourism and Leisure
Molchanova, Vidishcheva & Potapova	2018	Operating e-commercial market framework in terms of tourist destination	Espacios
Pikkemaat, Peters & Chan	2018	Needs, drivers and barriers of innovation: The case of an alpine community-model destination	Tourism Management Perspectives

(continued)

Table 2.1 (continued)

Authors	Year	Title	Source title
Shaheer, Insch & Carr	2018	Tourism destination boycotts–are they becoming a standard practise?	Tourism Recreation Research
Signorile, Larosa & Spiru	2018	Mobility as a service: A new model for sustainable mobility in tourism	Worldwide Hospitality and Tourism Themes
Almobaideen, Krayshan, Allan & Saadeh	2017	Internet of things: Geographical routing based on healthcare centres vicinity for mobile smart tourism destination	Technological Forecasting and Social Change
Bochert, Cismaru & Foris	2017	Connecting the members of generation Y to destination brands: A case study of the CUBIS project	Sustainability (Switzerland)
Chuang, Liu, Lu, Tseng, Lee & Chang	2017	The main paths of eTourism: Trends of managing tourism through Internet	Asia Pacific Journal of Tourism Research
Hardy, Hyslop, Booth, Robards, Aryal, Gretzel & Eccleston	2017	Tracking tourists' travel with smartphone-based GPS technology: A methodological discussion	Information Technology and Tourism
Kalbaska, Janowski, Estevez & Cantoni	2017	When digital government matters for tourism: A stakeholder analysis	Information Technology and Tourism
Kim & Kim	2017	The role of mobile technology in tourism: Patents, articles, news, and mobile tour app reviews	Sustainability (Switzerland)
Kuščer, Mihalič & Pechlaner	2017	Innovation, sustainable tourism, and environments in mountain destination development: A comparative analysis of Austria, Slovenia, and Switzerland	Journal of Sustainable Tourism
Li, Robinson & Oriade	2017	Destination marketing: The use of technology since the millennium	Journal of Destination Marketing and Management
Shao, Chang & Morrison	2017	How can big data support smart scenic area management? An analysis of travel blogs on Huashan	Sustainability (Switzerland)
Vishnevskaya, Klimova, Slinkova & Glumova	2017	The influence of virtual information spaces on tourism development	Espacios
Almobaideen, Allan & Saadeh	2016	Smart archaeological tourism: Contention, convenience, and accessibility in the context of cloud-centric IoT	Mediterranean Archaeology and Archaeometry

(continued)

Table 2.1 (continued)

Authors	Year	Title	Source title
Buonincontri & Micera	2016	The experience co-creation in smart tourism destinations: A multiple case analysis of European destinations	Information Technology and Tourism
Huang, Backman, Backman & Chang	2016	Exploring the implications of virtual reality technology in tourism marketing: An integrated research framework	International Journal of Tourism Research
Jamhawi & Hajahjah	2016	It-innovation and technologies transfer to heritage sites: The case of Madaba, Jordan	Mediterranean Archaeology and Archaeometry
Kaur	2016	Customer interface in spiritual tourism via "synaptic crm gap": An integrative technology-based conceptual model for relationship marketing	Journal of Relationship Marketing
Meehan, Lunney, Curran & McCaughey	2016	Aggregating social media data with temporal and environmental context for recommendation in a mobile tour guide system	Journal of Hospitality and Tourism Technology
Montaño & Ivanova	2016	Towards a new local sustainable development model for a consolidated tourist destination: The case of Los Cabos, Mexico	International Journal of Sustainable Development and Planning
Scott, Hall & Gössling	2016	A report on the Paris climate change agreement and its implications for tourism: Why we will always have Paris	Journal of Sustainable Tourism
Stipanovic & Rudan	2016	Tourism product club in generating the value chain	Polish Journal of Management Studies
Tussyadiah	2016	The influence of innovativeness on on-site smartphone use among American travelers: Implications for context-based push marketing	Journal of Travel and Tourism Marketing
Couture, Arcand, Sénécal & Ouellet	2015	The influence of tourism innovativeness on online consumer behaviour	Journal of Travel Research
Gretzel, Sigala, Xiang & Koo	2015	Smart tourism: Foundations and developments	Electronic Markets
Hjalager	2015	100 Innovations that transformed tourism	Journal of Travel Research
Palacios-Marqués, Merigó & Soto-Acosta	2015	Online social networks as an enabler of innovation in organisations	Management Decision

(continued)

Table 2.1 (continued)

Authors	Year	Title	Source title
Ali & Frew	2014a	ICT and sustainable tourism development: An innovative perspective	Journal of Hospitality and Tourism Technology
Ali & Frew	2014b	Technology innovation and applications in sustainable destination development	Information Technology and Tourism
Alizhonivna	2014	Standardisation of the sites of tourist destinations in Ukraine as a means of implementation of the Internet technologies on the local territorial level of the tourist area	Asian Social Science
Davidson & Keup	2014	The use of Web 2.0 as a marketing tool by the European Convention Bureaux	Scandinavian Journal of Hospitality and Tourism
Krizaj, Brodnik & Bukovec	2014	A tool for measurement of innovation newness and adoption in tourism firms	International Journal of Tourism Research
Pantano & Corvello	2014	Tourists' acceptance of advanced technology-based innovations for promoting arts and culture	International Journal of Technology Management
dos Santos	2014	Destination mutant brands: An empirical perspective of tourism professionals and host community to create Mutant Brand Cape Verde	Journal of Destination Marketing and Management
Scott & Frew	2014	Adoption of information and communications technology (ICT) by in-trip leisure tourists	Advances in Culture, Tourism and Hospitality Research
Chiabai, Paskaleva & Lombardi	2013	e-Participation model for sustainable cultural tourism management: A bottom-up approach	International Journal of Tourism Research
Neuts, Romão, Nijkamp & van Leeuwen	2013	Digital destinations in the tourist sector: A path model for the impact of e-services on tourist expenditures in Amsterdam	Letters in Spatial and Resource Sciences
Peeters	2013	Developing a long-term global tourism transport model using a behavioural approach: Implications for sustainable tourism policy making	Journal of Sustainable Tourism
Hjalager & Flagestad	2012	Innovations in well-being tourism in the Nordic countries	Current Issues in Tourism
McCabe, Sharples & Foster	2012	Stakeholder engagement in the design of scenarios of technology-enhanced tourism services	Tourism Management Perspectives

(continued)

Table 2.1 (continued)

Authors	Year	Title	Source title
Munar	2012	Social media strategies and destination management	Scandinavian Journal of Hospitality and Tourism
Neuhofer, Buhalis & Ladkin	2012	Conceptualising technology enhanced destination experiences	Journal of Destination Marketing and Management
Kuusik, Tiru, Ahas & Varblane	2011	Innovation in destination marketing: The use of passive mobile positioning for the segmentation of repeat visitors in Estonia	Baltic Journal of Management
Munar	2011	Tourist-created content: Rethinking destination branding	International Journal of Culture, Tourism and Hospitality Research
Moghimehfar & Nasr-Esfahani	2011	Decisive factors in medical tourism destination choice: A case study of Isfahan, Iran and fertility treatments	Tourism Management
Paskaleva, Cooper & Azorín	2011	Soft factors in integrating innovation in advanced e-services	International Journal of Services, Technology and Management
Peres, Correia & Moital	2011	The indicators of intention to adopt mobile electronic tourist guides	Journal of Hospitality and Tourism Technology
Reino, Frew & Albacete-Sáez	2011	ICT adoption and development: Issues in rural accommodation	Journal of Hospitality and Tourism Technology
Fuchs, Höpken, Föger & Kunz	2010	E-business readiness, intensity, and impact: An Austrian destination management organisation study	Journal of Travel Research
Hjalager	2010	A review of innovation research in tourism	Tourism Management
Jiricka, Salak, Eder, Arnberger & Pröbstl	2010	Energetic tourism: Exploring the experience quality of renewable energies as a new sustainable tourism market	WIT Transactions on Ecology and the Environment
Hyun, Lee, & Hu	2009	Mobile-mediated virtual experience in tourism: Concept, typology, and applications	Journal of Vacation Marketing
Navickas & Malakauskaite	2009	The possibilities for the identification and evaluation of tourism sector competitiveness factors	Engineering Economics
Ramona, Pirvu & Nanu	2009	A tourism marketing and brand management perspective in Romania	Metalurgia International

(continued)

Table 2.1 (continued)

Authors	Year	Title	Source title
Racherla, Hu & Hyun	2008	Exploring the role of innovative technologies in building a knowledge-based destination	Current Issues in Tourism
Lindroth, Ritalahti & Soisalon-Soininen	2007	Creative tourism in destination development	Tourism Review
Buhalis & O'Connor	2005	Information communication technology revolutionising tourism	Tourism Recreation Research
Gooroochurn & Sugiyarto	2005	Competitiveness indicators in the travel and tourism industry	Tourism Economics
Yaw	2005	Cleaner technologies for sustainable tourism: Caribbean case studies	Journal of Cleaner Production
Stamboulis & Skayannis	2003	Innovation strategies and technology for experience-based tourism	Tourism Management
Mansfeld	2002	Reinventing a destination through a "network" designed destination management system: The case of the rural north of Israel	Tourism
Gretzel, Yuan & Fesenmaier	2000	Preparing for the new economy: Advertising strategies and change in destination marketing organisations	Journal of Travel Research
Buhalis	1998	Strategic use of information technologies in the tourism industry	Tourism Management

of information technologies in the tourism industry", for which there are 436 author citations; third, Stamboulis and Skayannis's (2003) paper, "Innovation strategies and technology for experience-based tourism" with 274 author citations; and fourth, Gretzel et al.'s (2015) paper, "Smart tourism: Foundations and developments" with 258 author citations (Fig. 2.2).

Analysis of the journals in which there is research on innovation, technology, and tourism destinations revealed 47 journals (Table 2.2). They cover not only the fields of tourism and hospitality research but also the areas of general economics, management, innovation, engineering, and metallurgy, such as the *European Journal of Management and Business Economics*; *International Journal of Innovative Technology and Exploring Engineering*; *Mediterranean Archaeology and Archaeometry*; *Engineering Economics*; and *Metalurgia International.*

Despite the wide variety, the leading journals that register a significant number of published articles on destination innovation are (Table 2.2): *Journal of Sustainable Tourism* (six documents, 6.9%), *Journal of Destination Marketing and Management, Journal of Hospitality and Tourism Technology, Journal of Travel Research*

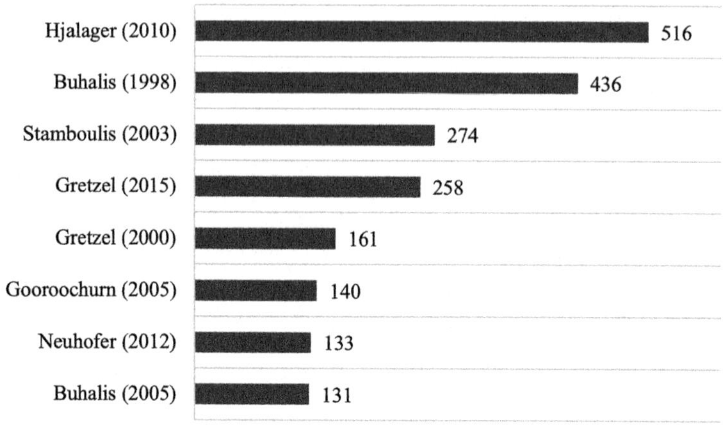

Fig. 2.2 The number of author citations for the article

and *Sustainability (Switzerland)* (five documents, 5.75%), *Information Technology and Tourism* and *Tourism Management* (four documents, 4.6%).

Interesting findings emerged considering the trends of the leading journals to publish articles on innovation in tourism destinations during different years. The *Journal of Sustainable Tourism*, *Sustainability (Switzerland)*, and the *Journal of Destination Marketing and Management*, presented a strong incremental effect of publishing. The *Journal of Hospitality and Tourism Technology* and *Information Technology and Tourism* manifest a gradual impact of publishing; finally, the *Journal of Travel Research and Tourism Management* is decremental in publishing on this topic.

Table 2.3 summarises the conceptual map of the forms of technology-driven innovation of the destination, considering:

1. *Five actors involved*: Tourists, firms, local community, destination management organisations (DMOs), institutions, and political actors.
2. Social capital as a *platform nurturing innovation.*
3. *Five forms of destination innovation*: Smart tourism, experience co-creation, e-participative governance, social innovation, and creative tourism.

This systematic literature review maps and discusses theoretical advances in technology-driven innovation in tourism destinations, contributing to the broader destination innovation debate. Far from being comprehensive in the complex and contextual phenomenon of technology-driven innovation of the destination, this literature review represents a preliminary synthetic depiction of possible drivers and forms of innovation, which the literature detected as presenting challenges, limitations, and future scenarios. It points towards new avenues for academics, policymakers, and destination managers and invites us to explore and advance theoretical research on human-technology interaction in the destination and its challenges.

Table 2.2 Article distribution across journals

Journal	No. of studies	%
Journal of Sustainable Tourism	6	6,9
Journal of Destination Marketing and Management	5	5,75
Journal of Hospitality and Tourism Technology	5	5,75
Journal of Travel Research	5	5,75
Sustainability (Switzerland)	5	5,75
Information Technology and Tourism	4	4,6
Tourism Management	4	4,6
International Journal of Tourism Research	3	3,44
Tourism Recreation Research	3	3,44
Tourism Review	3	3,44
Asia Pacific Journal of Tourism Research	2	2,3
Current Issues in Tourism	2	2,3
Espacios	2	2,3
International Journal of Innovative Technology and Exploring Engineering	2	2,3
Mediterranean Archaeology and Archaeometry	2	2,3
Scandinavian Journal of Hospitality and Tourism	2	2,3
Tourism Management Perspectives	2	2,3
Advances in Culture, Tourism and Hospitality Research	1	1,15
African Journal of Hospitality, Tourism and Leisure	1	1,15
African Journal of Science, Technology, Innovation and Development	1	1,15
ANATOLIA	1	1,15
Applied Sciences (Switzerland)	1	1,15
Asian Social Science	1	1,15
Baltic Journal of Management	1	1,15
Electronic Markets	1	1,15
Engineering Economics	1	1,15
European Journal of Management and Business Economics	1	1,15
International Journal of Culture, Tourism and Hospitality Research	1	1,15
International Journal of Scientific and Technology Research	1	1,15
International Journal of Services, Technology and Management	1	1,15
International Journal of Sustainable Development and Planning	1	1,15
International Journal of Technology Management	1	1,15
Journal of Cleaner Production	1	1,15
Journal of Relationship Marketing	1	1,15
Journal of Service Management	1	1,15
Journal of Travel and Tourism Marketing	1	1,15
Journal of Urban Technology	1	1,15
Journal of Vacation Marketing	1	1,15
Letters in Spatial and Resource Sciences	1	1,15
Management Decision	1	1,15
Metalurgia International	1	1,15

(continued)

Table 2.2 (continued)

Journal	No. of studies	%
Polish Journal of Management Studies	1	1,15
Technological Forecasting and Social Change	1	1,15
Tourism	1	1,15
Tourism Economics	1	1,15
WIT Transactions on Ecology and the Environment	1	1,15
Worldwide Hospitality and Tourism Themes	1	1,15
Total	87	100

2.6.1 Actors Involved in Destination Innovation

Interesting findings emerge from the analysis regarding the role of diverse actors involved in destination innovation processes (Table 2.4): the demand side (i.e. tourists) is most frequently the focus of research (in 78 articles, 89.65%). They confirm the difference between innovation in tourism (and services) and the manufacturing industry and allow us to underline how tourists—defined as prosumers (Chandler and Chen 2015)—are one of the key drivers of destination innovation.

Innovation in destinations is also supply-driven, leveraging top-down, and/or bottom-up processes. As our findings demonstrate, bottom-up processes nurture destination innovation: firms are important drivers (in 38 articles, 43.67%), and local communities contribute to destination (in 34 articles, 39.08%). These findings are coherent with the theoretical framework of local systems of innovation, amongst which destinations can be included (Camisón and Monfort-Mir 2012; Gomezelj 2016; Hjalager 2010; Novelli et al. 2006; Rodriguez et al. 2014; Svensson et al. 2005; Trunfio and Campana 2019).

DMOs are the main drivers of top-down destination innovation (in 31 articles, 35.63%) associated with institutions (in 28 articles, 32.18%).

The literature review unveils the relevant and pervasive role of the tourist in technology-driven innovation (89.65% of articles). Tourists exploit technological platforms and become active innovators during the design of creative and personalised content—such as travel information search, destination decisions, and travel product booking (Yoo et al. 2017)—driving the supply system towards satisfaction of their own experiences (Buhalis 2019; Buhalis et al. 2019; Koo et al. 2017; Lalicic and Dickinger 2019). Destination management research can cross-fertilise with the holistic innovation paradigm of user motivation (Norman and Verganti 2014), in which imaginative, cognitive, and affective elements create new spaces for tourist-driven innovations.

Besides this, innovative combinations between the physical and the virtual (on-site and online) spaces in experience co-creation represent possible areas of academic investigation. Tourists explore and exploit technological platforms to

Table 2.3 Drivers and innovations in the destination

Author	Year	Destination actors					Platforms		Innovations
		Tourists	Firms	Local community	DMOs	Institutions / political actors	ICTs	Social capital	
Barandiarán, Restrepo, and Luna	2019	X		X	X	X	X	X	E-participative governance
Buhalis	2019	X	X				X		Experience co-creation
Buhalis, Harwood, Bogicevic, Viglia, Beldona, and Hofacker	2019	X	X	X			X	X	Experience co-creation
Bulti, Ray, and Bhuyan	2019	X	X				X	X	Smart tourism
Chung, Tyan, and Lee	2019	X	X				X		Experience co-creation
Garcia-Sanchez, Siles, and Vazquez-Méndez	2019	X					X		Smart tourism
Gomez-Oliva, Alvarado-Uribe, Parra-Merono, and Jara	2019	X		X		X	X		Smart tourism
Lam-Gonzalez, Leon and de Leon	2019	X					X		Smart tourism
Liu and Nijkamp	2019	X	X	X		X	X	X	Social innovation
Maestro and Dumlao	2019	X		X		X	X		Smart tourism
Muliawaty, Alamsyah, and Loupias	2019	X	X	X		X	X		E-participative governance
Nilsson	2019	X		X		X	X		Smart tourism
Rastrollo-Horrillo and Rivero Diaz	2019	X	X	X		X	X	X	Social innovation
Sanz-Ibanez, Lozano, and Anton Clavé	2019		X		X	X	X		Smart tourism
Tshipala, Coetzee, and Potgieter	2019	X		X			X		Smart tourism
Weaver and Moyle	2019	X					X		Smart tourism
Xu, Nash, and Whitmarsh	2019	X		X		X	X		Smart tourism

(continued)

Table 2.3 (continued)

Author	Year	Destination actors					Platforms		Innovations
		Tourists	Firms	Local community	DMOs	Institutions / political actors	ICTs	Social capital	
Çakar, Kalbaska, Inanir, Şahin, and Ören	2018	X			X	X	X		Smart tourism
da Costa Liberato, Alén-González, and de Azevedo Liberato	2018a	X		X	X	X	X	X	Smart tourism
da Costa Liberato, Alén-González, and de Azevedo Liberato	2018b	X		X			X	X	Smart tourism
Ghaderi, Hatamifar, and Henderson	2018	X		X	X		X		Smart tourism
Lodeiro-Santiago, Santos-González, Caballero-Gil, Caballero-Gil, and Herrera-Priano	2018	X					X		Smart tourism
Mofokeng and Matima	2018	X	X				X		Experience co-creation
Molchanova, Vidishcheva, and Potapova	2018	X	X				X		Experience co-creation
Pikkemaat, Peters, and Chan	2018	X	X	X	X	X	X		Smart tourism
Shaheer, Insch, and Carr	2018	X					X		Social innovation
Signorile, Larosa, and Spiru	2018	X		X		X	X		Smart tourism
Almobaideen, Krayshan, Allan, and Saadeh	2017	X					X		Smart tourism
Bochert, Cismaru, and Foris	2017	X	X	X	X		X		Experience co-creation
Chuang, Liu, Lu, Tseng, Lee, and Chang	2017	X	X				X		Experience co-creation
Hardy, Hyslop, Booth, Robards, Aryal, Gretzel, and Eccleston	2017	X					X		Smart tourism
Kalbaska, Janowski, Estevez, and Cantoni	2017	X		X	X	X	X		E-participative governance

Kim and Kim	2017	X	X	X		X	X	X	Smart tourism
Kuščer, Mihalič, and Pechlaner	2017	X	X	X	X	X	X		Smart tourism
Li, Robinson, and Oriade	2017	X			X		X		Experience co-creation
Shao, Chang, and Morrison	2017	X					X		Smart tourism
Vishnevskaya, Klimova, Slinkova, and Glumova	2017	X					X		Experience co-creation
Almobaideen, Allan, and Saadeh	2016	X					X		Smart tourism
Buonincontri and Micera	2016	X	X		X	X	X	X	Experience co-creation
Huang, Backman, Backman, and Chang	2016	X		X			X		Experience co-creation
Jamhawi and Hajahjah	2016	X		X			X		Smart tourism
Kaur	2016	X	X			X	X	X	Experience co-creation
Meehan, Lunney, Curran, and McCaughey	2016	X					X		Experience co-creation
Montaño and Ivanova	2016		X				X		E-participative governance
Scott, Hall, and Gössling	2016		X			X	X		Smart tourism
Stipanović and Rudan	2016	X	X		X		X		Smart tourism
Tussyadiah	2016	X					X		Smart tourism
Couture, Arcand, Sénécal, and Ouellet	2015	X	X				X		Experience co-creation
Gretzel, Sigala, Xiang, and Koo	2015	X	X	X		X	X		Smart tourism
Hjalager	2015	X	X				X		Smart tourism
Palacios-Marqués, Merigó, and Soto-Acosta	2015	X	X				X		Experience co-creation
Ali and Frew	2014a	X		X	X		X		Smart tourism

(continued)

Table 2.3 (continued)

Author	Year	Destination actors					Platforms		Innovations
		Tourists	Firms	Local community	DMOs	Institutions / political actors	ICTs	Social capital	
Ali and Frew	2014b	X	X	X	X		X		Smart tourism
Alizhonivna	2014	X					X		Experience co-creation
Davidson and Keup	2014	X	X	X	X		X		Experience co-creation
Krizaj, Brodnik, and Bukovec	2014	X	X				X		Smart tourism
Pantano and Corvello	2014	X					X		Smart tourism
dos Santos	2014	X		X	X		X		Experience co-creation
Scott and Frew	2014	X					X		Smart tourism
Chiabai, Paskaleva, and Lombardi	2013	X		X		X	X		E-participative governance
Neuts, Romão, Nijkamp, and van Leeuwen	2013	X					X		Smart tourism
Peeters	2013					X	X		Smart tourism
Hjalager and Flagestad	2012	X	X		X		X		Smart tourism
McCabe, Sharples, and Foster	2012	X		X			X		E-participative governance
Munar	2012	X	X		X	X	X		Experience co-creation
Neuhofer, Buhalis, and Ladkin	2012	X			X		X		Experience co-creation
Kuusik, Tiru, Ahas, and Varblane	2011	X			X		X		Smart tourism
Munar	2011	X			X		X		Experience co-creation
Moghimehfar and Nasr-Esfahani	2011	X					X		Smart tourism

Paskaleva, Cooper, and Azorín	2011	X		X	X		X		E-participative governance
Peres, Correia, and Moital	2011	X		X			X		Smart tourism
Reino, Frew, and Albacete-Sáez	2011		X				X		Smart tourism
Fuchs, Höpken, Föger, and Kunz	2010	X	X		X		X		Smart tourism
Hjalager	2010	X	X		X	X	X		Smart tourism
Jiricka, Salak, Eder, Arnberger, and Pröbstl	2010	X				X	X		Smart tourism
Hyun, Lee, and Hu	2009	X			X		X		Experience co-creation
Navickas and Malakauskaite	2009				X		X		Smart tourism
Ramona, Pirvu, and Nanu	2009	X			X		X		Experience co-creation
Racherla, Hu, and Hyun	2008	X	X	X		X	X	X	Smart tourism
Lindroth, Ritalahti, and Soisalon-Soininen	2007	X	X	X	X		X		Creative tourism
Buhalis and O'Connor	2005	X	X				X		Smart tourism
Yaw	2005	X	X			X	X		Smart tourism
Gooroochurn and Sugiyarto	2005	X		X		X	X		Smart tourism
Stamboulis and Skayannis	2003	X		X			X		Experience co-creation
Mansfeld	2002		X		X		X		E-participative governance
Gretzel, Yuan, and Fesenmaier	2000	X			X		X		Smart tourism
Buhalis	1998	X	X		X		X		Smart tourism
Total		78	38	34	31	28	87	11	
%		89,65%	43,67%	39,08%	35,63%	32,18%	100%	12,64%	

Table 2.4 Actors involved in destination innovation

Type of actors	No. of actors in 87 studies	% on 87 studies
Tourist	78	89,65
Firm	38	43,67
Local community	34	39,08
DMO	31	35,63
Institution/political actors	28	32,18

become active innovators during the diverse phases of the journey, designing creative and personalised experience content, and co-creating their own experience in the destination. How smart technologies can facilitate, constrain, or transform the tourism experience co-creation (also in the online experience or in the post-experience phase) and influencing actors' future behaviour remains a flourishing area of investigation.

Diverse research questions emerge, e.g. Does technology bring tourists closer or further away from their destination? Do virtual and online experiences replace the physical destination visit? Will local authenticity and encounters with the local community continue to play a relevant role in the future of the destination? How and how much does smart tourism redefine tourist and destination actor interaction? What future is there for destination co-creation?

Looking at the other destination stakeholders (firms, 43.67%; local communities, 39.08%; DMOs, 35.63%; institutions and political actors, 32.18%), the analysis confirms a growing number of studies in which demand and supply perspectives have been combined and identifies the presence of top-down and bottom-up processes of destination innovation. Technology drives hybrid governance models, combining diverse internal and external actors and enhancing e-participation in destination co-creation.

2.6.2 *ICT-Based Tools in Destination Innovation*

ICT is an inclusive term summarising all the tools that store, share, or manipulate knowledge with digital support. They help to resolve or anticipate emerging environmental or social issues in the destination; mitigate tourism's negative impacts, achieve sustainable destination development; and frame new destination innovative scenarios (Ali and Frew 2014a, b; Montaño and Ivanova 2016).

Each ICT-based tool can be considered as a destination support platform in which various actors can interact, monitor, assist and/or integrate resources, skills, and knowledge during managerial decision-making and/or actions coherent with the environmental, socio-cultural, and economic destination aspects (Ali and Frew 2014a, b). However, they require an accurate level of technology usability and acceptance (Lee 2015) and relevant investments (Burton-Jones and Hubona 2006; Fuchs et al. 2010; Leung et al. 2015).

Table 2.5 uses Ali and Frew's (2014a, b) works to build an overview article on ICT-based tools from a destination, tourist, and business perspective, classifying the remaining 85 articles of our research.

2.6.3 Drivers and Forms of Destination Innovation

Smart tourism is the most common in the research (51 articles, 58.62%), amongst the five forms of destination innovation (Table 2.6). Experience co-creation is receiving incremental attention in destination research (24 articles, 27.58%), followed by e-participative governance (eight articles, 9.2%), social innovation (three articles, 3.44%), and creative tourism (one article, 1.15%).

Although social capital assumes a critical role in driving transition processes during technology adoption and acceptance in the destination (Grzegorczyk 2019; Lang and Ramírez 2017; Lee 2015; Trunfio and Campana 2019; Yan and Guan 2018), it remains an underdeveloped topic in destination innovation (in 11 articles, 12.64%). All forms of innovation deal with specific actors' interaction and the role of ICTs and social capital in the destination (Fig. 2.3).

Smart tourism mainly focuses on tourists (46), followed by the local community (19), firms, institutions and political actors, and DMOs (18). Experience co-creation articles focus on the relationship between tourists and firms (14), followed by DMOs (9), the local community (6), and institutions and political actors (4).

The three novel topics of research in technology-driven innovation—e-participation/e-government, social innovation, and creative tourism—focus entirely on tourists, combined with the key actors. Only three articles, one-participation/e-government, mention firms' role in the destination; the social innovation articles completely ignore the role of DMOs, whilst the creative tourism paper ignores the role of both institutions/political actors and social capital as a support platform in the destination. These areas thus present possible future areas for research on the role of DMOs and the institutions/political actors in social innovation and creative tourism driven by technology.

2.7 Smart Tourism

The research recognises an overarching vision of technology-driven innovation in the destination, building on the broader picture of smart tourism, which sheds light on the combination of technology, business, and society. Smart technologies interconnect diverse destination stakeholders (including tourists), enhancing experience co-creation and, at the same time, improving the destination's quality of life and sustainable development. This picture of the academic investigation, in which

Table 2.5 ICT-based tools in destination innovation

ICTs-based tools	Description	No. of articles	% on 85 articles	Authors
Community informatics	Communication channels to engage destination stakeholders in online social communities and platforms to co-create multimedia content projects, destination brand building, and sharing information	21	20,70	Buhalis (2019); Barandiarán et al. (2019); Muliawaty et al. (2019); Sanz-Ibáñez et al. (2019); Shaheer et al. (2018); Bochert et al. (2017); Kalbaska et al. (2017); Buonincontri and Micera (2016); Montaño and Ivanova (2016); Stipanovic and Rudan (2016); Palacios-Marqués et al. (2015); dos Santos (2014); Chiabai et al. (2013); McCabe et al. (2012); Munar (2012); Neuhofer et al. (2012); Munar (2011); Hjalager (2010); Ramona et al. (2009); Lindroth et al. (2007); Gretzel et al. (2000);
Tourism information system	Assistant tools to support the destination manager in decision-making during the use of business-critical information in the destination	16	18,82	Buhalis et al. (2019); Bulti et al. (2019); Garcia-Sanchez, Siles and Vazquez-Méndez (2019); Rastrollo-Horrillo and Rivero Diaz (2019); Weaver and Moyle (2019); da Costa Liberato et al. (2018); Molchanova et al. (2018); Pikkemaat et al. (2018); Kaur (2016); Couture et al. (2015); Gretzel et al. (2015); Paskaleva et al. (2011); Fuchs et al. (2010); Racherla et al. (2008); Buhalis (1998);
Location-based services	Tools based on location-sensitive information to influence tourists'choices in a sustainable way in the destination	11	12,94	Liu and Nijkamp (2019); Maestro and Dumlao (2019); Ghaderi et al. (2018); Lodeiro-Santiago et al. (2018); Kim and Kim (2017); Almobaideen et al. (2016); Meehan et al. (2016); Tussyadiah

(continued)

Table 2.5 (continued)

ICTs-based tools	Description	No. of articles	% on 85 articles	Authors
				(2016); Scott and Frew (2014); Moghimehfar and Nasr-Esfahani (2011); Peres et al. (2011);
Virtual tourism	Virtual immersion tools to stimulate alternative experiences in cultural heritage or other points of interest minimising congestion practices in the destination	8	9,41	Mofokeng and Matima (2018); Li et al. (2017); Vishnevskaya et al. (2017); Huang et al. (2016); Jamhawi and Hajahjah (2016); Pantano and Corvello (2014); Neuts et al. (2013); Hyun et al. (2009);
Environment management information systems	Integration tools to manage several environmental information sources during decision-making in destination development	7	8,23	Xu et al. (2019); Kuščer et al. (2017); Shao et al. (2017); Hjalager and Flagestad (2012); Navickas and Malakauskaite (2009); Buhalis and O'Connor (2005); Gooroochurn and Sugiyarto (2005);
Destination management system	Distribution tools of accessible tourism information to promote stakeholder engagement based on destination platforms	6	7,05	Gomez-Oliva et al. (2019); Çakar et al. (2018); Chuang et al. (2017); Alizhonivna (2014); Davidson and Keup (2014); Stamboulis and Skayannis (2003);
Geographical information system	Sustainable monitoring tools to build future indicators, projects, and models of destination development	4	4,71	Tshipala et al. (2019); Almobaideen et al. (2017); Krizaj et al. (2014); Reino et al. (2011);
Intelligent transport system	Mobility and accessibility tools to provide travelling options in the destination with new sustainable models	4	4,71	Nilsson (2019); Signorile et al. (2018); Hjalager (2015); Peeters (2013);
Carbon calculator	Monitoring tools of carbon emissions to define correct policies of energy consumed and a reduction in environmental impact in the destination	3	3,52	Scott et al. (2016); Jiricka et al. (2010); Yaw (2005)
Global positioning system	Satellite tracking tools to identify tourist spatial-	2	2,35	Hardy et al. (2017); Kuusik et al. (2011);

(continued)

Table 2.5 (continued)

ICTs-based tools	Description	No. of articles	% on 85 articles	Authors
	temporal distribution with tourists' smartphone use in the destination			
Computer simulation	Analysis and simulation tools of destination data to validate future systems management, strategies, and investments	1	1,17	Mansfeld (2002);
Economic impact analysis software	Tools based on institutional software to relieve the economic impacts of tourism in the destination	1	1,17	Chung et al. (2019);
Weather, climate, and ocean change forecasting software	Monitoring tools to forecast critical events based on weather, climate, and ocean risks and changes in destination	1	1,17	Lam-González et al. (2019);
Total		85	100	

Table 2.6 Forms of destination innovation

Innovation	No. of studies	%
Smart tourism	51	58,62
Experience co-creation	24	27,58
E-participative governance	8	9,2
Social innovation	3	3,44
Creative tourism	1	1,15
Total	87	100

technology enhances interaction and involvement, provides new areas of inquiry to frame and conduct theoretical research on human–technology interaction in destination co-creation, shaping political agendas and destination management practices.

Our literature identifies smart tourism as an increasing area of investigation, with 51 articles (58.62%). Several authors explored the smart innovation phenomenon as combining technological, social, and economic aspects of the destination domain and shaping political agendas and academic research (Buhalis and O'Connor 2005; Fuchs et al. 2010; Gooroochurn and Sugiyarto 2005; Gretzel et al. 2000; Hjalager 2010; Hjalager and Flagestad 2012; Jiricka et al. 2010; Krizaj et al. 2014; Kuusik et al. 2011; Moghimehfar and Nasr-Esfahani 2011; Navickas and Malakauskaite 2009; Neuts et al. 2013; Pantano and Corvello 2014; Peeters 2013; Peres et al. 2011; Racherla et al. 2008; Reino et al. 2011; Scott and Frew 2014; Yaw 2005).

Gretzel et al.'s (2015) seminal work—"Smart tourism: Foundations and developments", has clarified and sharpened the definition of "smart" tourism, identifying it as the combination of three critical components: smart destination, smart business ecosystem, and smart experience. Smartness enhances the multifunctional use of

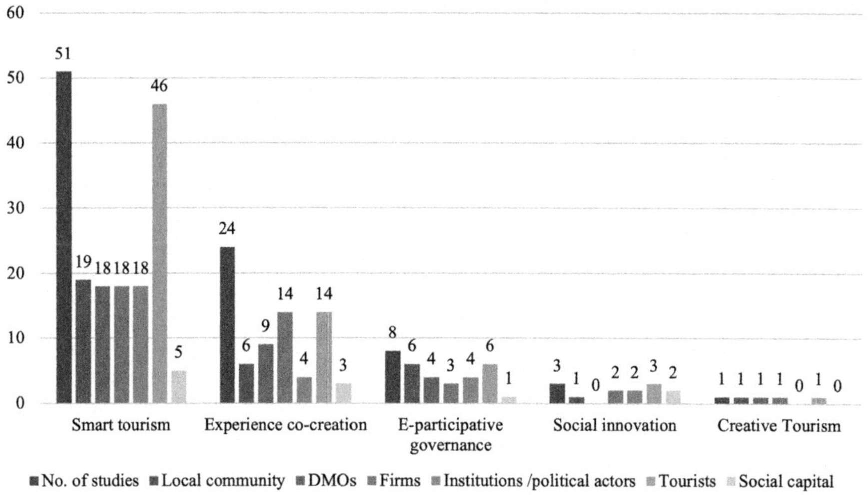

Fig. 2.3 Technology-based innovation: Drivers and forms of innovation

ICTs—the Internet of Things, wearables, mobile devices, virtual and augmented reality, cloud computing, services based on the user's location, and recommendation systems—to generate high levels of stakeholder interconnection and synchronisation (da Costa Liberato et al. 2018; Gretzel et al. 2015).

Table 2.7 adapts Gretzel et al.'s (2015) smart tourism components to classify the 50 remaining articles in the smart tourism innovation phenomenon.

Smart tourism integrates technological, social, and business components that interconnect tourism stakeholders (Buhalis and Amaranggana 2013; Yoo et al. 2017) and open spaces for the co-production and co-creation of experiences (Vargo and Lush 2013), overcoming the traditional reductionist perspective linked to the use of ICT tools.

Furthermore, smart tourism creates opportunities to improve the quality of life in tourism destinations and enhance sustainable stakeholder engagement (Kim and Kim 2017; Stipanovic and Rudan 2016). Destination governance plays a central role in creating the conditions for sustainable, innovative development by anticipating the needs of tourists and the local community and facilitating stakeholder engagement in value co-creation processes (Gretzel et al. 2015; Trunfio and Della Lucia 2019).

Table 2.7 Components of smart tourism

Smart tourism	No. of articles	% on 50 articles	Authors
The smart destination—Like smart cities—Considers the use of interconnected technological tools to link destination stakeholders, enhancing opportunities for communication, collaboration, mutual support, and knowledge sharing to enable innovation at multiple levels. It is a digital place in which hard and soft technologies foster stakeholder interaction and integration, creating new opportunities for improving the quality of life based on efficiency-efficacy, sustainability, accessibility, public and private amenities, and human resource valorisation.	29	58	Bulti et al. (2019); Gomez-Oliva et al. (2019); Lam-González et al. (2019); Maestro and Dumlao (2019); Nilsson (2019); Ghaderi et al. (2018); Lodeiro-Santiago et al. (2018); Pikkemaat et al. (2018); Signorile et al. (2018); Almobaideen et al. (2017); Hardy et al. (2017); Kim and Kim (2017); Kuščer et al. (2017); Shao et al. (2017); Almobaideen et al. (2016); Jamhawi and Hajahjah (2016); Tussyadiah (2016); Ali and Frew (2014a, b); Pantano and Corvello (2014); Scott and Frew (2014); Neuts et al. (2013); Peeters (2013); Kuusik et al. (2011); Moghimehfar and Nasr-Esfahani (2011); Peres et al. (2011); Jiricka et al. (2010); Racherla et al. (2008); Yaw (2005);
The smart business ecosystem involves complex connections and relations between public and private actors, and local communities and tourists, to create ecosystems in which sharing knowledge is used to anticipate the needs of tourists and the local community, facilitating stakeholder engagement in value co-creation processes.	15	30	Sanz-Ibáñez et al. (2019); Tshipala et al. (2019); Xu et al. (2019); da Costa Liberato et al. (2018); Scott et al. (2016); Stipanovic and Rudan (2016); Krizaj et al. (2014); Hjalager and Flagestad (2012); Reino et al. (2011); Fuchs et al. (2010); Hjalager (2010); Navickas and Malakauskaite (2009); Gooroochurn and Sugiyarto (2005); Gretzel et al. (2000); Buhalis (1998);
The smart experience enabled diverse social and technological technologies to interconnect tourism stakeholders and open spaces for the co-production and co-creation of experience.	6	12	Lam-González et al. (2019); Weaver and Moyle (2019); Çakar et al. (2018); da Costa Liberato, Alén-González and de Azevedo Liberato (2018b); Chuang et al. (2017); Hjalager (2015); Buhalis and O'Connor (2005);
Total	50	100	

2.8 The Smart Tourism Mindset Combining Technology- and Social-Driven Innovation

The literature has criticised the isolated technological approach in destination innovation and the empty rhetoric of smart tourism initiatives, calling for the introduction of "liquidity into the smart tourism ecosystem" [..] so "data and human capital flow freely and that partnerships can be formed easily so that the smart tourism ecosystem remains agile as well as permeable"(Gretzel 2022, p. 8). New approaches have been proposed to involve actors in destination innovation, reframing DMOs' role in smart processes and increase the "smartness" of stakeholders (Gretzel 2022; Gretzel and Collier de Mendonça 2019; Trunfio and Campana 2019).

Pitfalls and constraints emerged, interpreting technologies as a disruptive innovation in destination management and neglecting the role of stakeholders in understanding destination smartness (Gelter et al. 2022; Gretzel 2021; Gretzel 2022). Although smart technologies promote several opportunities, their use can generate restraints in human–technology interaction. They can be overload, technostress, sickness, and difficulty to use (Stankov and Filimonau 2019; Trunfio et al. 2022). Besides limitations in technology acceptance and adoption would compromise actors' participation in the innovation processes of the destination (Lee 2015; Pérez-Luño et al. 2011).

Literature called for overcoming the myopic vision of technology-driven innovation and building on a smart tourism mindset, as a constructive and collective re-imagining of changes allows us to reframe the innovation path of the destination (Gretzel 2021).

A smart tourism mindset allows us to frame a holistic and overarching destination innovation perspective and builds on the broader vision of smartness in tourism where smart tourism ideas and values enhance engagement and develop an attitude to anticipate, drive, and implement pervasive innovation which combines technological and social changes. A virtuous process of socially embedded and inclusive innovation in the destination can emerge by leveraging human–technology interaction through continuous investments in ICT infrastructure and social capital. Combining technology- and social-driven innovations reshapes the social context. It reinforces actors' connections, knowledge sharing, and dissemination, contributing to innovation through stakeholder engagement and multi-level governance. Investing in social-driven innovations stimulates destination actors in the change process, promoting ICTs' acceptance and adoption (van der Have and Rubalcaba 2016; Lee 2015; Pérez-Luño et al. 2011; Trunfio and Campana 2020). Social capital can create the conditions for collaboration, trust, and knowledge sharing, which, in turn, encourage the acceptance, use, and diffusion of ICTs, and their effective and widespread technology-driven innovation, including experience co-creation and sustainable destination development (Lee 2015; Trunfio and Campana 2019).

The complementary roles of social capital and ICT infrastructure create conditions that facilitate human–technology interactions, nurturing alternative and unexplored forms of destination innovation and overcoming the traditional limitations of the technology-driven approach (Lee 2015; Pérez-Luño et al. 2011; Trunfio and Campana 2019; van der Have and Rubalcaba 2016). As a result, innovative destination models emerge, combining smartness, cultural heritage, creativity, social innovation, and entrepreneurship, reshaping the role of each local actor, creating new business opportunities, and re-vitalising the cultural values of the local community (Carvalho and van Winden 2017; Mengi et al. 2017; Pique et al. 2019; Ponto and Inkinen 2019; Trunfio and Della Lucia 2019).

Evolutionary processes, combining demand and supply perspectives and top-down and bottom-up approaches, shape destination innovation, opening new avenues for academics, policymakers, and destination managers to explore and advance theoretical research on human–technology interaction and its challenges. Further investigations can underpin the ICTs' role in designing new scenarios for advances in destination management research considering (e.g.): re-engineering processes and networking building between diverse actors, including tourists, which enhance collective innovativeness and social changes; incubators of creative forms of innovation, which drive new experiences and unusual destination models; marketing intelligence sources which exploit technologies 4.0 (e.g. the Internet of Things, big data analytics, artificial intelligence, augmented reality, virtual reality, and mixed reality) (Ali and Frew 2014a, b; Buhalis and Law 2008; Sigala 2018).

Interpreting the systematic literature review, three forms of technology-driven innovation that lever on some social-driven innovation aspects (e.g. actors, networking, and trust) can be investigated: e-participative governance, experience co-creation, and social innovation.

2.9 E-Participative Governance

E-participative governance represents a challenging area of investigation in destination management and governance. It combines traditional governance models, based on the top-down approach, with digital stakeholder engagement (Sigala and Marinidis 2012; Trunfio and Della Lucia 2019).

Multi-level and shared governance models have been extensively researched in destination management. Most of the literature affirms the shift from top-down governance models towards forms of destination governance in which destination stakeholders participate in the decision-making process (Beritelli 2011; Bramwell and Lane 2011; Go and Trunfio 2011; Go et al. 2013; Laws et al. 2011; Ruhanen et al. 2010).

The evolution of destinations has now resulted in hybrid models (Montaño and Ivanova 2016; Rodriguez et al. 2014), in which ICTs accelerate stakeholder engagement (Mansfeld 2002; Paskaleva et al. 2011) and create spaces for e-participative destination governance models (Sigala and Marinidis 2012; Trunfio and Della Lucia 2019).

ICTs and digital platforms provide vital digital spaces for stakeholder engagement and shared governance (involving DMOs, firms, institutions, local communities, and tourists) in destination development. Only eight of the articles (9.2%) in our literature review investigated e-participative governance processes as drivers of destination innovation (Barandiarán et al. 2019; Chiabai et al. 2013; Kalbaska et al. 2017; Mansfeld 2002; McCabe et al. 2012; Montaño and Ivanova 2016; Muliawaty et al. 2019; Paskaleva et al. 2011). They analyse stakeholder participation forms in network-based tourism products and conclude that networks result from the failure of top-down governance in destination management (Barandiarán et al. 2019;

Muliawaty et al. 2019). Hybrid models and bottom-up processes facilitate democratic stakeholder participation in decision-making and raise local awareness of tourism value (Chiabai et al. 2013; Kalbaska et al. 2017; McCabe et al. 2012).

E-participation encourages knowledge sharing, strengthens multi-level coordination and collaboration amongst stakeholders, and enhances stakeholder engagement in problem solving, introducing new public services (Barandiarán et al. 2019; Montaño and Ivanova 2016; Muliawaty et al. 2019). ICT tools—such as e-forums, e-blogs, and focus groups—facilitate access to the digital spaces in which tourism development projects and diverse forms of e-participation in decision-making are proposed and discussed (Chiabai et al. 2013).

2.10 Experience Co-creation

Despite experience being an already well-researched topic in tourism studies—as an evolution of the traditional experience economy (Pine and Gilmore 1999; Prahalad and Ramaswamy 2004)—the ICT-based experience co-creation stream of research is relatively novel and underdeveloped (24 articles, 27.58%). ICTs, e-tourism, mobile technologies, social media, and gamification play a relevant role in building experience co-creation, and increasing interactions between tourists and destination stakeholders. This phenomenon is explored in Stamboulis and Skayannis's (2003) seminal work—with 274 authors' citations—followed by several other articles (Buhalis et al. 2019; Buonincontri and Micera 2016; Neuhofer et al. 2012).

The experience co-creation paradigm in the digital era represents the evolution of the traditional experience economy (Pine and Gilmore 1999; Prahalad and Ramaswamy 2004): leveraging technologies, tourists make informed decisions and reduce their risk and fear, becoming prosumers during their experience in the destination (pre-travel, on-site, and post-travel) (Neuhofer et al. 2012). Experience co-creation in the different phases can be redesigned and recreated by technological tools, allowing virtual or physical connections with destination suppliers and stakeholders, other tourists, virtual communities, and family, friends, and peers (Alizhonivna 2014; Chung et al. 2019; Huang et al. 2016; Hyun et al. 2009; Li et al. 2017; Mofokeng and Matima 2018; Molchanova et al. 2018; Ramona et al. 2009; Vishnevskaya et al. 2017).

Internet-based tools promote active and authentic tourist participation and increase their absorption-immersion levels by giving them new opportunities in real-time interactive channels to co-create, co-market, co-produce, and co-design the destination, creating a new social experience ecosystem (Bochert et al. 2017; Buhalis 2019; Buhalis et al. 2019; Buhalis and Sinarta 2019; Couture et al. 2015; Meehan et al. 2016; Munar 2011, 2012; Neuhofer et al. 2012; Palacios-Marqués et al. 2015; dos Santos 2014; Stamboulis and Skayannis 2003).

Destination managers build on new technologies to develop a personalised and contextualised experience creation (Buhalis 2019; Buhalis et al. 2019; Buhalis and Sinarta 2019; Chuang et al. 2017; Davidson and Keup 2014). They empower productive dynamic interactions and instant gratification, thus generating value for tourists (Buhalis et al. 2019; Buonincontri and Micera 2016; Kaur 2016; Stamboulis and Skayannis 2003).

2.11 Social Innovation

Although social innovation is a newly emerging research stream that can involve any innovation in destination, very limited literature contributions have been identified in the technology-driven innovation field of research (3.4%).

Several authors have explored this phenomenon with a multidisciplinary literature perspective, defining an ambiguous state based on a continuous process of knowledge accumulation (Choi and Majumdar 2015; van der Have and Rubalcaba 2016).

Systematic literature created a complete thematic overview of the established body of knowledge about social innovation and conceptualised the innovation phenomenon as a new social service practice based on new creative ideas (van der Have and Rubalcaba 2016). It allows the revitalisation of the traditional social structure of certain local contexts, in which several urban or rural problems of marginalisation (e.g. declines in public spending, low quality of citizens' life, empowerment imbalance of local actors, injustice, and underprivileged) are embedded.

Social innovation redefines the destination innovation patterns engaging unusual stakeholders at multiple levels, such as local communities, non-profit and non-government organisations (NGOs). It problematises and mobilises social changes in specific networks of resources, power relations, and communication skills (Neumeier 2017; Oliveira and Breda-Vázquez 2012; Rastrollo-Horrillo and Rivero Díaz 2019). Social capital facilitates actors' integration and goal alignment. It promotes collaboration, the dissemination of learning, and the inspiration or replication of successful social projects (Lang and Ramírez 2017; Liu and Nijkamp 2019; Rastrollo-Horrillo and Rivero Díaz 2019) involving contexts other than those that are high-tech (Yan and Guan 2018).

Digital platforms and other ICT-based tools improve knowledge-sharing practices (Grzegorczyk 2019), amplifying the reshaping phenomenon of local contexts with new creative value (Koo et al. 2017). The destination becomes a non-profit incubator and/or accelerator of social changes driven by bottom-up approaches (Edwards-Schachter and Wallace 2017; van der Have and Rubalcaba 2016; Rutten and Boekema 2007; Shaheer et al. 2018).

Technology can remove the traditional limitations of intensive mediation between social innovators and lawmakers, but it cannot improve territorial cohesion, cultural diversity, and tolerance (Grzegorczyk 2019). It develops homogeneous connections that promote urban or rural policies of social innovation legitimisation (Koo et al. 2017; Neumeier 2017; Oliveira and Breda-Vázquez 2012; Rastrollo-Horrillo and Rivero Díaz 2019).

2.12 Creative Tourism

Only one of the articles in the literature review interpreted creative tourism as adopting the lens of technology-driven destination innovation. This research topic is undeveloped, considering technology potentialities in facilitating and reshaping forms of creativity in tourism destinations.

Authors considered creative tourism an alternative concept to social innovation (Lindroth et al. 2007). They emphasised how specific local cultural contexts attract externally talented people—such as artists, writers, scientists, entertainers, actors, and designers—to promote urban or rural projects of the creative economy at the level of cities and regions (Della Lucia and Trunfio 2018; Oliveira and Breda-Vázquez 2012). Building on cultural heritage theory, the authors analysed how cultural regeneration strategies of destination stakeholders help tourism destinations rethink local development creatively. The research demonstrated how—when the local community is proactive—the destination can enhance creativity and innovativeness—by exploiting social capital to use ICT infrastructures—to facilitate knowledge sharing between diverse actors and innovative human capital management approaches.

References

Adams R, Jeanrenaud S, Bessant J, Denyer D, Overy P (2016) Sustainability-oriented innovation: a systematic review. Int J Manag Rev 18:180–205

Aarstad J, Ness H, Haugland SA (2015) Innovation, uncertainty, and inter-firm shortcut ties in a tourism destination context. Tour Manag 48:354–361

Adler PS, Kwon S-W (2002) Social capital: prospects for a new concept. Acad Manag Rev 27 (1):17–40

Ali A, Frew AJ (2014a) ICT and sustainable tourism development: an innovative perspective. J Hosp Tour Technol 5:2–16

Ali A, Frew AJ (2014b) Technology innovation and applications in sustainable destination development. Inf Tech Tour 14:265–290

Alizhonivna ZK (2014) Standardisation of the sites of tourist destinations in Ukraine as a means of implementation of the internet technologies on the local territorial level of the tourist area. Asian Soc Sci 10:273

Almobaideen W, Allan M, Saadeh M (2016) Smart archaeological tourism: contention, convenience and accessibility in the context of cloud-centric IoT. Mediter Archaeol Archaeom 16: 227–236

Almobaideen W, Krayshan R, Allan M, Saadeh M (2017) Internet of things: geographical routing based on healthcare centers vicinity for mobile smart tourism destination. Technol Forecast Soc Chang 123:342–350

Asheim B, Coenen L, Vang J (2007) Face-to-face, buzz, and knowledge bases: sociospatial implications for learning, innovation, and innovation policy. Environ Plann C: Gov Policy 25 (5):655–670

Baggio R, Cooper C (2010) Knowledge transfer in a tourism destination: the effects of a network structure. Serv Ind J 30:1–15

Barandiarán X, Restrepo N, Luna Á (2019) Collaborative governance in tourism: lessons from Etorkizuna Eraikiz in the Basque Country, Spain. Tour Rev 74:902–914
Beritelli P (2011) Cooperation among prominent actors in a tourist destination. Ann Tour Res 38: 607–629
Beritelli P, Bieger T (2014) From destination governance to destination leadership – defining and exploring the significance with the help of a systemic perspective. Tour Rev 69(1):25–46
Beritelli P, Bieger T, Laesser C (2016) The new frontiers of destination management: applying variable geometry as a function-based approach. J Travel Res 53(4):403–417
Bochert R, Cismaru L, Foris D (2017) Connecting the members of generation y to destination brands: a case study of the CUBIS project. Sustainability 9:1–20
Boekema F, Morgan K, Bakkers S, Rutten R (2000) Knowledge, innovation and economic growth. Edward Elgar Publishing
Boes K, Buhalis D, Inversini A (2015) Conceptualising smart tourism destination dimensions. In: Information and communication technologies in tourism, pp 391–403
Bramwell B, Lane B (2011) Critical research on the governance of tourism and sustainability. J Sustain Tour 19:411–421
Buhalis D (2019) Technology in tourism-from information communication technologies to eTourism and smart tourism towards ambient intelligence tourism: a perspective article. Tour Rev 75:267
Buhalis D (1998) Strategic use of information technologies in the tourism industry. Tour Manag 19: 409–421
Buhalis D, Amaranggana A (2015) Smart tourism destinations enhancing tourism experience through personalisation of services. In: Information and communication Technologies in Tourism 2015. Springer, Cham, pp 377–389
Buhalis D, Amaranggana A (2013) Smart tourism destinations. In: Information and communication Technologies in Tourism 2014. Springer, Cham, pp 553–564
Buhalis D, Harwood T, Bogicevic V, Viglia G, Beldona S, Hofacker C (2019) Technological disruptions in services: lessons from tourism and hospitality. J Serv Manag 30:484–506
Buhalis D, Law R (2008) Progress in information technology and tourism management: 20 years on and 10 years after the internet-the state of eTourism research. Tour Manag 29:609–623
Buhalis D, O'Connor P (2005) Information communication technology revolutionising tourism. Tour Recreat Res 30:7–16
Buhalis D, Sinarta Y (2019) Real-time co-creation and nowness service: lessons from tourism and hospitality. J Travel Tour Mark 36:563–582
Bulti AG, Ray A, Bhuyan P (2019) Smart tourism system architecture design using the internet of everything (IOE) over cloud platform. Int J Innov Tech Explor Engg 8:421–426
Buonincontri P, Micera R (2016) The experience co-creation in smart tourism destinations: a multiple case analysis of European destinations. Inf Tech Tour 16:285–315
Burton-Jones A, Hubona GS (2006) The mediation of external variables in the technology acceptance model. Inf Manag 43:706–717
Cabiddu F, Lui TW, Piccoli G (2013) Managing value co-creation in the tourism industry. Ann Tour Res 42:86–107
Çakar K, Kalbaska N, Inanir A, Şahin Ören T (2018) eVisa's impacts on travel and tourism: the case of Turkey. J Hosp Tour Technol 9:14–32
Camisón C, Monfort-Mir VM (2012) Measuring innovation in tourism from the Schumpeterian and the dynamic-capabilities perspectives. Tour Manag 33:776–789
Carrillo FJ (2014) 'What', 'knowledge-based' stands for? A position paper. Int J Knowl Based Dev 5(4):402–421
Carvalho L, van Winden W (2017) Planned knowledge locations in cities: studying emergence and change. Int J Knowl Based Dev 8(1):47
Centobelli P, Ndou V (2019) Managing customer knowledge through the use of big data analytics in tourism research. Curr Issue Tour 22(15):1862–1882

Chandler J, Chen S (2015) Prosumer motivations in service experiences. J Serv Theory Pract 25: 220–239
Chiabai A, Paskaleva K, Lombardi P (2013) E-Partecipation model for sustainable cultural tourism management: a bottom-up approach. Int J Tour Res 15:35–51
Choi N, Majumdar S (2015) Social innovation: towards a conceptualisation. In: Technology and innovation for social change. Springer, New Delhi, pp 7–34
Chuang TC, Liu JS, Lu LYY, Tseng FM, Lee Y, Chang CT (2017) The main paths of eTourism: trends of managing tourism through internet. Asia Pacific J Tour Res 22:213–231
Chung N, Tyan I, Lee SJ (2019) Eco-innovative museums and visitors' perceptions of corporate social responsibility. Sustainability 11:5744
Coleman JS (1988) Social capital in the creation of human capital. Am J Sociol 94:S95–S120
Corfu A, Kastenholz E (2005) The opportunities and limitations of the internet in providing a quality tourist experience. J Qual Assur Hosp Tour 6:77–88
Couture A, Arcand M, Sénécal S, Ouellet JF (2015) The influence of tourism innovativeness on online consumer behavior. J Travel Res 54:66–79
da Costa Liberato PM, Alén-González E, de Azevedo Liberato DFV (2018) Smart tourism destination triggers consumer experience: the case of Porto. Eur J Manag Bus Econ 27:6–25
Davidson R, Keup M (2014) The use of web 2.0 as a marketing tool by European convention Bureaux. Scand J Hosp Tour 14:234–254
Del Chiappa G, Baggio R (2015) Knowledge transfer in smart tourism destinations: analysing the effects of a network structure. J Destin Mark Manag 4:145–150
Del Vecchio P, Passiante G (2017) Is tourism a driver for smart specialization? Evidence from Apulia, an Italian region with a tourism vocation. J Destin Mark Manag 6(3):163–165
Del Vecchio P, Mele G, Ndou V, Secundo G (2018) Creating value from social big data: implications for smart tourism destinations. Inf Process Manag 54(5):847–860
Della Lucia M, Trunfio M (2018) The role of the private actor in cultural regeneration: hybridising cultural heritage with creativity in the city. Cities 82:35–44
dos Santos ERM (2014) Destination mutant brands: an empirical perspective of tourism professionals and host community to create mutant brand Cape Verde. J Destin Mark Manag 3:68–78
Edwards-Schachter M, Wallace ML (2017) 'Shaken, but not stirred': sixty years of defining social innovation. Technol Forecast Soc Chang 119:64–79
Edwards-Schachter M (2018) The nature and variety of innovation. Int J Innov Stud 2(2):65–79
Fagerberg J (2017) Innovation policy: rationales, lessons and challenges. J Econ Surv 31(2): 497e512
Fagerberg J (2018) Mobilizing innovation for sustainability transitions: a comment on transformative innovation policy. Res Policy 47:1568–1576
Fagerberg J, Verspagen B (2009) Innovation studies: the emerging structure of a new scientific field. Res Policy 38(2):218e233
Femenia-Serra F, Neuhofer B, Ivars-Baidal JA (2019) Towards a conceptualisation of smart tourists and their role within the smart destination scenario. Serv Ind J 39:109–133
Flagestad A, Hope CA, Svensson B, Nordin S (2005) The tourist destination: a local innovation system? The creation of a model. In: Keller P, Bieger T (eds) Innovation in tourism-creating customer value. AIEST, Brainerd, MN, pp 245–259
Fuchs M, Höpken W, Föger A, Kunz M (2010) E-business readiness, intensity, and impact: an Austrian destination management organisation study. J Travel Res 49:165–178
Gelter J, Fuch M, Lexhagen M (2022) Making sense of smart tourism destinations: a qualitative text analysis from Sweden. J Destin Mark Manag 23:1–12
Ghaderi Z, Hatamifar P, Henderson JC (2018) Destination selection by smart tourists: the case of Isfahan, Iran. Asia Pacific J Tour Res 23:385–394
Go FM, Trunfio M (2011) E-services governance in public and private sectors: a destination management organisation perspective. In: Information technology and innovation trends in organisations. Physica-Verlag HD, pp 11–19

Go FM, Trunfio M, Della Lucia M (2013) Social capital and governance for sustainable rural development. Stud Agri Econ 115:104–110

Gomez-Oliva A, Alvarado-Uribe J, Parra-Meroño MC, Jara AJ (2019) Transforming communication channels to the co-creation and diffusion of intangible heritage in smart tourism destination: creation and testing in Ceutí (Spain). Sustainability 11:1–30

Gomezelj DO (2016) A systematic review of research on innovation in hospitality and tourism. Int J Contemp Hosp Manag 28:516–558

Gooroochurn N, Sugiyarto G (2005) Competitiveness indicators in the travel and tourism industry. Tour Econ 11:25–43

Gretzel U (2021) Conceptualizing the smart tourism mindset: fostering utopian thinking in smart tourism development. J Smart Tour 1(1):3–8

Gretzel U (2022) The smart DMO: a new step in the digital transformation of destination management organisations. Euro J Tour Res 30:302

Gretzel U, Collier de Mendonça M (2019) Smart destination brands: semiotic analysis of visual and verbal signs. Int J Tourism Cities 5(4):560–580

Gretzel U, Sigala M, Xiang Z, Koo C (2015) Smart tourism: foundations and developments. Electron Mark 25:179–188

Gretzel U, Yuan YL, Fesenmaier DR (2000) Preparing for the new economy: advertising strategies and change in destination marketing organisations. J Travel Res 39:146–156

Grzegorczyk M (2019) The role of culture-moderated social capital in technology transfer – Insights from Asia and America. Technol Forecast Soc Chang 143:132–141

Hardy A, Hyslop S, Booth K, Robards B, Aryal J, Gretzel U, Eccleston R (2017) Tracking tourists' travel with smartphone-based GPS technology: a methodological discussion. Inf Technol Tour 17:255–274

Hjalager AM (2015) 100 innovations that transformed tourism. J Travel Res 54:3–21

Hjalager AM (2010) A review of innovation research in tourism. Tour Manag 31:1–12

Hjalager AM, Flagestad A (2012) Innovations in Well-being tourism in the Nordic countries. Curr Issue Tour 15:725–740

Hoarau H, Kline C (2014) Science and industry: sharing knowledge for innovation. Ann Tour Res 46:44–61

Hossain M (2018) Frugal innovation: a review and research agenda. J Clean Prod 182:926–936

Huang YC, Backman KF, Backman SJ, Chang LL (2016) Exploring the implications of virtual reality technology in tourism marketing: an integrated research framework. Int J Tour Res 18: 116–128

Hyun MY, Lee S, Hu C (2009) Mobile-mediated virtual experience in tourism: concept, typology and applications. J Vacat Mark 15:149–164

Inkpen AC, Tsang EW (2005) Social capital, networks, and knowledge transfer. Acad Manag Rev 30:146–165

Işık C, Aydın E, Dogru T, Rehman A, Sirakaya-Turk E, Karagöz D (2022) Innovation research in tourism and hospitality field: a bibliometric and visualization analysis. Sustainability 14:1–21

Ivars-Baidal JA, Celdrán-Bernabeu MA, Mazón JN, Perles-Ivars ÁF (2019) Smart destinations and the evolution of ICTs: a new scenario for destination management? Curr Issue Tour 22 (13):1581–1600

Jamhawi MM, Hajahjah ZA (2016) It-innovation and technologies transfer to heritage sites: the case of Madaba. Jordan Mediterr Archaeol Archaeom 16:41–46

Jiricka A, Salak B, Eder R, Arnberger A, Pröbstl U (2010) Energetic tourism: exploring the experience quality of renewable energies as a new sustainable tourism market. Trans Ecol Environ 139:55–68

Kahn KB (2018) Understanding innovation. Bus Horiz 61(3):453–460

Kalbaska N, Janowski T, Estevez E, Cantoni L (2017) When digital government matters for tourism: a stakeholder analysis Nadzeya. Inf Technol Tour 17:315–333

Kaur G (2016) Customer interface in spiritual tourism via "synaptic CRM gap": an integrative technology-based conceptual model for relationship marketing. J Relationship Market 15:326–343
Kim D, Kim S (2017) The role of mobile technology in tourism: patents, articles, news, and mobile tour app reviews. Sustainability 9:2082
Koo C, Tussyadiah IP, Hunter WC (2017) Special issue on how technology-enhanced tourism is transforming societies, cultures and economies. Technol Forecast Soc Chang 123:327–329
Krizaj D, Brodnik A, Bukovec B (2014) A tool for measurement of innovation newness and adoption in tourism firms. Int J Tour Res 16:113–125
Kuščer K, Mihalič T, Pechlaner H (2017) Innovation, sustainable tourism and environments in mountain destination development: a comparative analysis of Austria, Slovenia and Switzerland. J Sustain Tour 25:489–504
Kuusik A, Tiru M, Ahas R, Varblane U (2011) Innovation in destination marketing: the use of passive mobile positioning for the segmentation of repeat visitors in Estonia. Balt J Manag 6: 378–399
Lalicic L, Dickinger A (2019) An assessment of user-driven innovativeness in a mobile computing travel platform. Technol Forecast Soc Chang 144:233–241
Lam-González YE, León CJ, de León Ledesma J (2019) Assessing the effects of the climatic satisfaction on nautical tourists' on-site activities and expenditure decisions. J Destin Mark Manag 14:100372
Lang T, Ramírez R (2017) Building new social capital with scenario planning. Technol Forecast Soc Chang 124:51–65
Laws E, Richins H, Agrusa J, Scott N (2011) Tourist destination governance decision making: complexity, dynamics and influences. In: Tourist destination governance practice, theory and issues, pp 83–90
Lee BC (2015) The impact of social capital on tourism technology adoption for destination marketing. Curr Issue Tour 18:561–578
Leung D, Lo A, Hoc L, Fong N, Law R (2015) Applying the technology-organisation- environment framework to explore ICT initial and continued adoption: an exploratory study of an independent hotel in Hong Kong. Tour Recreat Res 40:391–406
Li SCH, Robinson P, Oriade A (2017) Destination marketing: the use of technology since the millennium. J Destin Mark Manag 6:95–102
Lindroth K, Ritalahti J, Soisalon-Soininen T (2007) Creative tourism in destination development. Tour Rev 62:53–58
Liu CH, Lin JY (2012) Social relationships and knowledge creation: the mediate of critical network position. Serv Ind J 32(9):1469–1488
Liu J, Nijkamp P (2019) Inbound tourism as a driving force for regional innovation: a spatial impact study on China. J Travel Res 58:594–607
Lodeiro-Santiago M, Santos-González I, Caballero-Gil C, Caballero-Gil P, Herrera-Priano F (2018) Novel guidance CPS based on the FatBeacon protocol. Appl Sci 8:647
Maestro NB, Dumlao MF (2019) Romblon islands into a smart tourism destination through point of interest recommender, augmented reality and near field communication: a proposal. Int J Innov Technol Exploring Engg 8:242–247
Mansfeld Y (2002) Reinventing a destination through a "network" designed destination management system-the case of the rural north of Israel. Tourism 50:349–359
Marasco A, De Martino M, Magnotti F, Morvillo A (2018) Collaborative innovation in tourism and hospitality: a systematic review of the literature. Int J Contemp Hosp Manag 30:2364–2395
Martínez-Pérez Á, Elche D, García-Villaverde PM (2019) From diversity of interorganizational relationships to radical innovation in tourism destination: the role of knowledge exploration. J Destin Mark Manag 11:80–88
McCabe S, Sharples M, Foster C (2012) Stakeholder engagement in the design of scenarios of technology-enhanced tourism services. Tour Manag Perspect 4:36–44

Meehan K, Lunney T, Curran K, McCaughey A (2016) Aggregating social media data with temporal and environmental context for recommendation in a mobile tour guide system. J Hosp Tour Technol 7:281–299
Mengi O, Drinkwater SBD, Öner AC, Velibeyoğlu K (2017) Place management of a creative city: the case of Izmir. Int J Hum Comput Stud 8:271–291
Mofokeng NEM, Matima TK (2018) Future tourism trends: virtual reality based tourism utilising distributed ledger technologies. African J Hosp Leisure, Sport Tour 7:1–14
Moghimehfar F, Nasr-Esfahani MH (2011) Decisive factors in medical tourism destination choice: a case study of Isfahan, Iran and fertility treatments. Tour Manag 32:1431–1434
Molchanova VS, Vidishcheva EV, Potapova II (2018) Operating e-commercial market framework in terms of tourist destination. Espacios 39:8–14
Montaño A, Ivanova A (2016) Towards a new local sustainable development model for a consolidated tourist destination: the case of Los Cabos, Mexico. Int J Sustain Dev Plan 11:138–146
Muliawaty L, Alamsyah K, Loupias HH (2019) The role of new public service model in developing tourist destination in Bandung city: a new paradigm for domestic tourism industry. Int J Sci Technol Res 8:322–326
Munar AM (2012) Social media strategies and destination management. Scand J Hosp Tour 12: 101–120
Munar AM (2011) Tourist-created content: rethinking destination branding. Int J Cult Tour Hosp Res 5:291–305
Nahapiet J, Ghoshal S (1998) Social capital, intellectual capital, and the organisational advantage: sociological and economic approaches to the analysis of social structure. Acad Manag Rev 23: 242–266
Narduzzo A, Volo S (2018) Tourism innovation: when interdependencies matter. Curr Issue Tour 21:735–741
Navickas V, Malakauskaite A (2009) The possibilities for the identification and evaluation of tourism sector competitiveness factors. Eng Econ 1:37–44
Nespolo D, Fachinelli AC, Fortes VMM, Milan GS, Camargo ME (2018) Knowledge-based development from the citizen's perspective: a study from southern Brazil. Int J Knowl Based Dev 9:6–22
Neuhofer B, Buhalis D, Ladkin A (2012) Conceptualising technology enhanced destination experiences. J Destin Mark Manag 1:36–46
Neumeier S (2017) Social innovation in rural development: identifying the key factors of success. Geogr J 183:34–46
Neuts B, Romão J, Nijkamp P, van Leeuwen E (2013) Digital destinations in the tourist sector: a path model for the impact of e-services on tourist expenditures in Amsterdam. Lett Spat Resour Sci 6:71–80
Nilsson JH (2019) Urban bicycle tourism: path dependencies and innovation in greater Copenhagen. J Sustain Tour 27:1648–1662
Norman DA, Verganti R (2014) Incremental and radical innovation: design research vs. technology and meaning change. Des Issues 30:78–96
Novelli M, Schmitz B, Spencer T (2006) Networks, clusters and innovation in tourism: a UK experience. Tour Manag 27:1141–1152
OECD (2005) "The measurement of scientific and technological activities: guidelines for collecting and interpreting innovation data: oslo manual, third edition" prepared by the working Party of National Experts on scientific and technology indicators. OECD, Paris
OECD (2017) Science, technology and industry scoreboard 2017. The digital transformation. OECD Publishing, Paris, France
Okubo Y (1997) Bibliometric indicators and analysis of research systems. OECD Science, Technology and Industry Working Papers. 1997/01.
Oliveira C, Breda-Vázquez I (2012) Creativity and social innovation: what can urban policies learn from sectoral experiences? Int J Urban Reg Res 36:522–538

Ozseker DB (2018) Towards a model of destination innovation process: an integrative review. Serv Ind J 39:206–228
Paget E, Dimanche F, Mounet JP (2010) A tourism innovation case. An actor-network approach. Ann Tour Res 37:828–847
Palacios-Marqués D, Merigó JM, Soto-Acosta P (2015) Online social networks as an enabler of innovation in organisations. Manag Decis 53:1906–1920
Pantano E, Corvello V (2014) Tourists' acceptance of advanced technology-based innovations for promoting arts and culture. Int J Technol Manag 64:3–16
Paskaleva K, Cooper I, Azorín JA (2011) Soft factors in integrating innovation in advanced e-services. Int J Serv Technol Manag 15:161–177
Peeters PM (2013) Developing a long-term global tourism transport model using a behavioural approach: implications for sustainable tourism policy making. J Sustain Tour 21:1049–1069
Peres R, Correia A, Moital M (2011) The indicators of intention to adopt mobile electronic tourist guides. J Hosp Tour Technol 2:120–138
Pérez-Luño A, Medina CC, Lavado AC, Rodríguez GC (2011) How social capital and knowledge affect innovation. J Bus Res 64:1369–1376
Pickering C, Byrne J (2014) The benefits of publishing systematic quantitative literature reviews for PhD candidates and other early-career researchers. High Educ Res Dev 33:534–548
Pike S, Page SJ (2014) Destination marketing organisations and destination marketing: a narrative analysis of the literature. Tour Manag 41:202–227
Pikkemaat B, Peters M, Chan CS (2018) Needs, drivers and barriers of innovation: the case of an alpine community-model destination. Tour Manag Perspect 25:53–63
Pine BJ, Gilmore JH (1999) The experience economy: work is theatre & every business a stage. Harvard Business Press
Pique MJ, Miralles F, Teixeira CS, Gaspar JV, Filho JRBR (2019) Application of the triple helix model in the revitalisation of cities: the case of Brazil. Int J Knowl Based Dev 10:43–74
Ponto H, Inkinen T (2019) Knowledge-based environments in the city: design and urban form in the Helsinki metropolitan area. Int J Knowl Based Dev 10(2):155–175
Prahalad CK, Ramaswamy V (2004) Co-creation experiences: the next practice in value creation. J Interact Mark 18:5–14
Putnam RD (1993) The prosperous community: social capital and public life. Am Prospect 13:35–42
Racherla P, Hu C, Hyun MY (2008) Exploring the role of innovative technologies in building a knowledge-based destination. Curr Issue Tour 11:407–428
Ramona G, Pirvu G, Nanu R (2009) A tourism marketing and brand management perspective in Romania. Metalurgia Int 14:140–142
Rastrollo-Horrillo MA, Rivero Díaz M (2019) Destination social capital and innovation in SMEs tourism firms: an empirical analysis in an adverse socio-economic context. J Sustain Tour 27: 1572–1590
Reino S, Frew AJ, Albacete-Sáez C (2011) ICT adoption and development: issues in rural accommodation. J Hosp Tour Technol 2:66–80
Rodriguez I, Williams AM, Hall CM (2014) Tourism innovation policy: implementation and outcomes. Ann Tour Res 49:76–93
Roper S, Love JH (2018) Knowledge context, learning and innovation: an integrating framework. Ind Innov 25:339–364
Ruhanen L, Scott N, Ritchie B, Tkaczynski A (2010) Governance: a review and synthesis of the literature. Tour Rev 65:4–16
Russo-Spena T, Tregua M, Bifulco F (2017) Searching through the jungle of innovation conceptualisations. System, network and ecosystem perspectives. J Serv Theory Pract 27: 977–1005
Rutten R, Boekema F (2007) Regional social capital: embeddedness, innovation networks and regional economic development. Technol Forecast Soc Chang 74:1834–1846

Sanz-Ibáñez C, Lozano S, Anton Clavé S (2019) Brokers in a destination's knowledge networks. J Destin Mark Manag 11:120–129
Sanzo-Perez MJ, Álvarez-González LI, Rey-García M (2015) How to encourage social innovations: a resource-based approach. Serv Ind J 35:430–447
Schuhbert A (2021) From network-pools to activated networks—a conceptual approach on absorptive capacities in a rural destination of Azerbaijan. J Inf Knowl Manag 20(02)
Schumpeter JA (1934) Change and the entrepreneur. Essays of JA Schumpeter
Scott D, Hall CM, Gössling S (2016) A report on the Paris climate change agreement and its implications for tourism: why we will always have Paris. J Sustain Tour 24:933–948
Scott MM, Frew AJ (2014) Adoption of information and communications technology (ICT) by in-trip leisure tourists. Adv Cult Tour Hosp Res 9:75–84
Shafiee S, Rajabzadeh Ghatari A, Hasanzadeh A, Jahanyan S (2019) Developing a model for sustainable smart tourism destinations: a systematic review. Tour Manag Perspect 31:287–300
Shaheer I, Insch A, Carr N (2018) Tourism destination boycotts–are they becoming a standard practise? Tour Recreat Res 43:129–132
Shao J, Chang X, Morrison AM (2017) How can big data support smart scenic area management? An analysis of travel blogs on Huashan. Sustainability 9:1–17
Sheehan L, Vargas-Sánchez A, Presenza A, Abbate T (2016) The use of intelligence in tourism destination management: an emerging role for DMOs. Int J Tour Res 18:549–557
Sigala M (2018) New technologies in tourism: from multi-disciplinary to anti-disciplinary advances and trajectories. Tour Manag Perspect 25:151–155
Sigala M, Marinidis D (2012) E-democracy and web 2.0: a framework enabling DMOs to engage stakeholders in collaborative destination management. Tour Anal 17:105–120
Signorile P, Larosa V, Spiru A (2018) Mobility as a service: a new model for sustainable mobility in tourism. Worldwide Hospitality and Tourism Themes 10:185–200
Stamboulis Y, Skayannis P (2003) Innovation strategies and technology for experience-based tourism. Tour Manag 24:35–43
Stankov U, Filimonau V (2019) Reviving calm technology in the e-tourism context. Serv Ind J 39: 343–360
Stankov U, Gretzel U (2020) Tourism 4.0 technologies and tourist experiences: a human-centered design perspective. Inf Technol Tour 22(3):477–488
Stipanovic C, Rudan E (2016) Tourism product club in generating the value chain. Pol J Manag Stud 14:214–224
Svensson B, Nordin S, Flagestad A (2005) A governance perspective on destination development-exploring partnerships, clusters and innovation systems. Tour Rev 60:32–37
Torraco RJ (2016) Writing integrative literature reviews: using the past and present to explore the future. Hum Resour Dev Rev 15:404–428
Tribe K (2007) Strategies of economic order: German economic discourse. Cambridge University Press
Trunfio M, Campana S (2019) Drivers and emerging innovations in knowledge-based destinations: towards a research agenda. J Destin Mark Manag 14:1–12
Trunfio M, Campana S (2020) Innovation in knowledge-based destination: technology-driven versus social-driven. Int J Knowl Based Dev 11(2):76–199
Trunfio M, Jung T, Campana S (2022) Mixed reality experiences in museums: exploring the impact of functional elements of the devices on visitors' immersive experiences and post-experience behaviours. Inf Manag 59:1–16
Trunfio M, Della Lucia M (2019) Co-creating value in destination management levering on stakeholder engagement value. e-Review of Tourism Research 16:195–204
Tuli SC, Hu R, Dare L (2019) Planning a global knowledge city: experience from Melbourne, Australia. Int J Knowl Based Dev 10(1):26–42
Rosca E, Arnold M, Bendul JC (2017) Business models for sustainable innovation–an empirical analysis of frugal products and services. J Clean Prod 162:S133–S145

Saito H, Ruhanen L (2017) Power in tourism stakeholder collaborations: power types and power holders. J Hosp Tour Manag 31:189–196
Tshipala N, Coetzee WJ, Potgieter M (2019) Sustainable indicators for adventure tourism destinations: a case of Waterval Boven. Afr J Sci Technol Innov Dev 11:589–595
Tussyadiah IP (2016) The influence of innovativeness on on-site smartphone use among American travelers: implications for context-based push marketing. J Travel Tour Mark 33:806–823
Van Assche K, Beunen R, Holm J, Lo M (2013) Social learning and innovation. Ice fishing communities on lake mille lacs. Land Use Policy 34:233–242
van der Have RP, Rubalcaba L (2016) Social innovation research: an emerging area of innovation studies? Res. Policy 45:1923–1935
Vargo SL, Lush RF (2013) Foundations of service dominant logic. In: Naples forum, doctoral consortium
Vishnevskaya EV, Klimova TB, Slinkova OK, Glumova YG (2017) The influence of virtual information spaces on tourism development. Espacios 38:1–5
Walman L (2016) A review of the literature on citation impact indicators. J Informet 10:365–391
Weaver DB, Moyle BD (2019) 'Tourist stupidity' as a basic characteristic of 'smart tourism': challenges for destination planning and management. Tour Recreat Res 44:387–391
When U, Evers J (2015) The social innovation potential of ICT-enabled citizen observatories to increase eParticipation in local flood risk management. Technol Soc 42:187–198
Xin S, Tribe J, Chambers D (2013) Conceptual research in tourism. Ann Tour Res 41:66–88
Xu F, Nash N, Whitmarsh L (2019) Big data or small data? A methodological review of sustainable tourism. J Sustain Tour 28:1–20
Yan Y, Guan JC (2018) Social capital, exploitative and exploratory innovations: the mediating roles of ego-network dynamics. Technol Forecast Soc Chang 126:244–258
Yang ECL, Khoo-Lattimore C, Arcodia C (2017) A systematic literature review of risk and gender research in tourism. Tour Manag 58:89–100
Yaw JF (2005) Cleaner technologies for sustainable tourism: Caribbean case studies. J Clean Prod 13:117–134
Yigitcanlar T, Lönnqvist A (2013) Benchmarking knowledge-based urban development performance: results from the international comparison of helsinki. Cities 31:357–369
Yoo CW, Goo J, Huang CD, Nam K, Woo M (2017) Improving travel decision support satisfaction with smart tourism technologies: a framework of tourist elaboration likelihood and self-efficacy. Technol Forecast Soc Chang 123:330–341
Yung R, Khoo-Lattimore C (2019) New realities: a systematic literature review on virtual reality and augmented reality in tourism research. Curr Issue Tour 22:2056–2081
Zehrer A, Raich F, Siller H, Tschiderer F (2014) Leadership networks in destinations. Tour Rev 69 (1):59–73

Chapter 3
Smart and Sustainable Destination Management: An Analytical Framework

3.1 Sustainable and Responsible Tourism: A Destination Management Perspective

Sustainable tourism has witnessed long-term and growing attention in tourism studies (Butler 1999). It refers to a set of distinct and multifaceted theoretical frameworks that attempt to address the complexity of the disequilibria that tourism activities and growth create. Besides the theoretical and practical limitations of the sustainability concept, the sustainable tourism agenda drew significant attention to the dark side that tourism has, like any other industry.

Since the 1970s, attention has been drawn to the effects of a growing concentration of visitors in certain tourism destinations characterised by high seasonality, with tourists' behaviour impacting the local quality of life and profoundly changing the physical environment (Rosenow and Pulsipher 1979). The perceived social costs of tourists' concentration (Pizam 1978) and residents' attitudes to crowds of tourists (Stokols 1972) were early concerns. These evolved into the broader sustainability agenda in tourism, which historically kept a significant focus on residents' and visitors' perceptions of the intense tourism presence within destinations (Teye et al. 2002; Yeh et al. 2012; Jin et al. 2016; Rasoolimanesh et al. 2017).

Notions like the carrying capacity were applied to measure the load capacity of destinations based on the number of visitors in a given time, in light of visitors' mobility and itineraries and behaviour (McCool and Lime 2001). Recent additions to the concept stressed the importance of a continuous definition and estimation of the visitation optimum, acknowledging the influence of time and space, constantly changing the carrying capacity (Pásková et al. 2021). Technological aspects have significant implications in mediating the tourist load over the destination, well presented by Beritelli (2019), criticising the operational value of the traditional destination lifecycle *à la* Butler.

Growing tourism burdens were coupled with the observation of a rise in resident–visitor conflicting relations (Ap 1992). More recently, the sustainable tourism debate

C. Pasquinelli, M. Trunfio, *Sustainability-oriented Innovation in Smart Tourism*, Tourism on the Verge, https://doi.org/10.1007/978-3-031-33677-5_3

catalysed much attention towards overtourism, whose negative impacts were discussed in relation to disparate destinations, from urban to rural, mountain, and coastal destinations (see below; Peeters et al. 2018; Milano et al. 2019; Pasquinelli and Trunfio 2020a).

Today, sustainability issues and their multifaceted nature are generally acknowledged and considered, at least in principle. Policymakers and destination managers have started reflecting on limiting, managing, and mitigating negative tourism impacts. The media's sustainability narrative (e.g. extensive and global attention was drawn to overtourism) may influence public opinion's understanding of tourism and its role in local development (Hall 2003) until there is the shaping of actions (Schweinsberg et al. 2017).

Until recent times, "tourism [was] represented as a Panglossian panacea for many (in some cases even all) development problems (as a source of revenue, ideas, employment, connection, and dynamism [. . .])" (Pasquinelli and Bellini 2017, p. 5). Tourism can be a crucial asset for local communities, a fast-growing industry that creates economic value from local resources (UNWTO 2012). The spending capacity of international visitors is a source of revenue for disparate sectors, including hospitality, transport, sport, retailing, and creative industries (Hedrick-Wong and Choong 2014; Tyrrell and Johnston 2006). In several circumstances, tourism revealed its capacity to bounce back.

Several territories and destinations turned to tourism to boost local development. Many cities hit by de-industrialisation undertook culture-led regeneration processes (Bailey et al. 2004; Comunian and Sacco 2006; O'Brian and Miles 2010), boosting tourism-based consumption economies (Gotham 2002). Originally understood as the *soft* alternative to the old *hard* industry in crisis, tourism growth has clearly manifested its hard impact on social, cultural, and environmental resources. Besides, an extensive debate framed critical aspects such as issues of placelessness (Evans 2003), empty "eventification" (Jakob 2013), and the impossibility of defining and transferring any "easy" tourism development recipe (Franklin 2016).

In the entirely different context of inner areas, small rural and mountain villages, peripheral areas in relation to the urban poles where essential services are located (i.e. education, rail transport, and health services), tourism development is often envisioned as the way towards local development, giving residents a chance to stay. The *Italian National Strategy of Inner Areas* revealed this widespread view since all pilot areas' strategic documents positioned tourism at the centre of their strategy (Andreoli et al. 2017).

In all these different contexts, the emphasis on tourism development's positive impacts and economic benefits is based—consciously or not—on the tourism-led growth hypothesis, according to which incoming visitors draw monetary flows to tourism destinations (Brida et al. 2014), boosting wealth and wellbeing. These monetary flows, however, have in some cases overshadowed "the conceptualisation of the tourist flows and their impacts" (Pasquinelli and Bellini 2017, p. 7). The sustainable tourism agenda has been critically addressing the tourism-led growth hypothesis and has discussed its implications. It conceptually framed the related issues yet rarely indicated pragmatic directions to address them (on this aspect, see below). The translation of theoretical advancements into robust practice and policy implications concerning—for instance—destinations' and their tourism models' vulnerability and the balance between tourism burdens and benefits for local stakeholders is seriously needed.

In this regard, sustainable tourism research has witnessed the centrality of the stakeholder viewpoint. An example is the early and constant attention to two crucial stakeholders, residents, and tourists, focusing on their perceptions of tourism impacts. The stakeholder perspective revealed the need to adopt a *micro-perspective* segmenting each stakeholder category into distinct sub-groups. Analysis of residents' perceptions of tourism impacts revealed both negative (e.g. Perdue et al. 1999; Stewart and Cole 2001; Fleishman et al. 2004; Riganti and Nijkamp 2008) and positive perceptions (Addis and Holbrook 2006; Neuts and Nijkamp 2012; Ward and Berno 2011; Kim et al. 2016), depending on personal characteristics (Stokols 1972) and individual involvement with tourism from an economic perspective (Andereck et al. 2005), directly and indirectly.

In a recent editorial, Cheer et al. (2021), discussing the relationship between tourism and justice, summarised two key dimensions explaining central sustainable tourism concerns. They drew attention to the global-local tourism system raising "critical questions of responsibility, control, governance and use of the local and global commons and related issues of planetary sustainability and the wellbeing of human and non-human others" (p. 284). Then, referring to the significant role of trans-national operators, such as hotel chains, cruise companies, transport carriers, and intermediaries, the authors highlighted the *multiple stakeholders* and their different interests as key motivations for the tourism systems' complexity and extreme fragmentation. These authors also mentioned the potentially disruptive impact of technologies and the rapid evolution of ICTs, suggesting the need to further reflect on the effective and realistic emerging opportunities and challenges.

On the one hand, the tourism management debate has focused on the local system to acknowledge the importance of dealing with local networks of actors that are shaping the tourism system. On the other, however, the need is to frame and manage the "global pipelines" (borrowing the expression from Bathelt et al. 2004) of the local destination. Actors, resources, and networks shaping the global pipelines are part of the destination, so responsibility, control, governance, and the use of resources (Cheer et al. 2021) should be framed accordingly.

The growing debate—coupled with the intensification and increasing visibility of tourism impacts—trickled down into a mobilisation of consensus, policies, and financial resources at supranational, national, and local levels, towards underpinning a sustainable turn in tourism development. The general point of agreement achieved by scholars, practitioners, and policymakers is that we need to *do something* to pursue a sustainable tourism path. Tourism systems are not self-regulating systems—like any other industry—creating the need to act and choose to address the unavoidable impacts of tourism growth, maximising the positive impacts (these being economic, social, cultural, or environmental) and minimising or removing the negative ones (these being economic, social, cultural, or environmental). That is, tourism is an unsustainable industry that, *only if managed*, may undertake "more" sustainable development models.

The need to "take action" or, in contrast, the observation of inaction is the *fil rouge* of the sustainability agenda in tourism studies. By adopting an extremely diverse theoretical lens of observation, tourism studies connected the tourism industry and its actors with local communities and the environment, putting future generations' interests at centre stage (Ruhanen et al. 2015). Numerous flaws and gaps persist in the sustainable tourism research agenda. Sustainability remains a fuzzy concept and, in tourism studies, tends to be myopic, needing to reflect on the many connections of local systems with broader territorial and global tourism systems (see above, Cheer et al. 2021).

Despite the acknowledgement of the need to take action against imbalances and disequilibria, the "rhetoric of balance" between economic, social, and environmental dimensions remains difficult to translate into practice (Garrod and Fyall 1998; Liu 2003). Applied research advancements are needed to give decision makers concrete support. For instance, the carrying capacity concept (see above) is hardly applicable and arbitrary to some extent. This is due to the hard definition of the evaluative criteria that, on the one hand, need to be objective and, on the other, need to mirror the *desired state of equilibrium* for the destination. This is generally hard to define, extremely subjective, and observer dependent (Lindberg et al. 1997).

Similar limitations do apply even considering concepts designed for planning and management purposes. An example is the acceptable change model (McCool 1994) and similar planning and managerial frameworks (McCool and Lime 2001). These are built on an ex ante definition of the tolerable tourism load for a destination. A key point that these approaches raised is the need for *stakeholder participation and collective processes to identify tolerance thresholds.* Subsequently and indirectly, these bring about tourism governance issues towards creating the preconditions for organisational dynamics at the destination.

The stakeholder perspective is remarkably the pivot of the sustainable tourism agenda. However, unavoidable limitations and constraints persist concerning sustainable tourism development *in practice* with regard to the impossibility of solving the puzzle of all the different stakeholders' individual interests, mindsets, and interpretations of the tourism-local development relationship, overcoming their potential conflicts (Dwyer 2018). "Taking action" means taking realistic actions (Reinhold et al. 2015) and still many scholars' efforts are needed to help practitioners in this direction.

Sustainable tourism and sustainable development concepts revealed an intrinsic weakness when attempting to turn sustainability into policies. The collective pursuit of sustainability, based on the interdependences, relations, and connections amongst stakeholders (internal and external, local and global, tourism and non-tourism operators), has often implied a lack of pragmatic answers, pushing decision makers and destination managers towards bottlenecks that are hardly overcome.

Therefore, another concept emerged in the constellation of the sustainable tourism agenda, which is *responsible tourism.* Even though sustainable tourism and responsible tourism are not always distinguished in practice and in the literature, responsible tourism was defined as a practice-oriented interpretation of sustainability goals to boost their adoption in daily decision-making and actions (Goodwin 2019; Burrai et al. 2019). Practice-oriented interpretation means that responsible tourism calls for the personal responsibility of tourism stakeholders, who are individually responsible for choosing and behaving responsibly towards sustainable tourism. As tourists, residents, or tourism firms, each stakeholder contributes individually to sustainable tourism. The individual does his/her part by contributing to more sustainable tourism outside any policy or collective framework for action.

The responsible tourism concept has been criticised from different perspectives. It was maintained that complex sustainability challenges are necessarily systematic, regarding a complex system that does not correspond to the sum of the single components. Accordingly, they cannot be addressed through individual "atomised" actions. The "feel-good" practices of individuals may not be enough and, especially, the single actor may miss the broader picture of sustainability, which may require the effective pursuit of a change of model in the entire tourism system (Burrai et al. 2019).

Let us also remind ourselves of Cheer et al.'s editorial (2021), which pointed to the global-local tourism system. They clearly suggested how a change of model oriented towards sustainability may not be enough if it is conceived only by looking inside the destination (starting from local stakeholders), whilst forgetting the intense connections of the destination with the global tourism system and related actors. Furthermore, empirical evidence has shown that individual actors tend to make only limited investments in responsible tourism management (the study focused on tourism firms). In contrast, governments' support makes a remarkable difference in triggering actors' efforts for a sustainable transition (Frey and George 2010).

3.1.1 The Link Between Sustainability and Competitiveness

Narrowing the focus on sustainable tourism and looking at the destination management literature, the negligible amount of scholarship that attempts to build conceptual models broadly framing the sustainable destination is surprising. Much attention has been devoted to discussing specific and partial aspects of sustainable destination management (e.g. planning, practices, and tools for visitor management). However, scholars' engagement with broader conceptualisations of a sustainable destination and sustainable destination management was limited.

The gap in the literature is fairly evident concerning the link between destination competitiveness, which is a cornerstone of destination management, and sustainability. In the 1990s and early 2000s, much attention was paid to the notion of destination competitiveness, in literature and in practice. The influence of New Public Management in the literature was evident and penetrated tourism studies and policy making. The concept of competitiveness—mirroring the debate on regional and place competitiveness—has been focused on the determinants and indicators of a destination's "success", that is the destination's capacity to achieve and maintain competitive positioning in the widening international tourism market.

Determinants of destination competitiveness were framed, such as "endowed" and "created" resources, a system of destination management in which both private and public actors play an active role, situational conditions, and market demand (Dwyer and Kim 2003). Similarly, the comparative advantage (resource endowment) and competitive advantage (resource deployment as a constructed advantage), destination management, policy, and planning were central to the seminal contribution by Ritchie and Crouch (2003) on destination competitiveness.

These contributions remain key references when attempting to bridge competitiveness and sustainability conceptually. Despite not explicitly referring to sustainability, Dwyer and Kim's model (Dwyer and Kim 2003) included socio-economic prosperity in their framework as one of the likely effects of the competitive destination, to be monitored through quality-of-life indicators. Destination management is the pivot for generating such effects, but the causal link between competitiveness and socio-economic prosperity was not fully explained. Socio-economic prosperity is part of the model and is assumed to be the outcome of a managerial process that creates the conditions for the reproduction of the destination over time and, so, for the reproduction of the factors making the destination competitive (e.g. natural assets as endowed resources). Sustainability (not explicitly mentioned in this study) seems to be understood as the durability of the adopted tourism destination competitiveness model.

Ritchie and Crouch (2003) explicitly juxtaposed competitiveness and sustainability in their book, *The competitive destination. A sustainable tourism perspective*, which represented the first and the most systemic attempt to combine destination competitiveness and sustainability. The long-term perspective on destination reproduction is also present in this study as well ("Throughout this chapter we have emphasised that the competitiveness of a tourist destination must be assessed from a long-term perspective", p. 29). The authors underlined the multidimensional nature of the tourism destination's strength, which is the capacity to keep competitive positioning over time, putting aside the environmental, economic, social, cultural, political, and technological dimensions.

Interestingly for the aims of this book, Ritchie and Crouch commented on the technological dimension, highlighting its recent inclusion amongst the determinants of destination competitiveness (the book was published in 2003). They suggested that this was "only a minor factor in determining the appeal of destinations [but] it has become a major factor in the promotion and distribution of the travel experience for an increasing number of destinations" (p. 5).

As the authors stated concerning competitiveness and sustainability:

> In summary, what makes a tourism destination truly competitive is its ability to increase tourism expenditure, to increasingly attract visitors while providing them with satisfying, memorable experiences, and to do so in a profitable way, *while* [emphasis added by the authors] enhancing the wellbeing of destination residents and preserving the natural capital of the destination for future generations (p. 2).

Sustainability and competitiveness may be read as two parallel yet interconnected dimensions, with the former guaranteeing the durability of the latter. Values and philosophy inspiring destination management (Crouch 2011) integrate the sustainability vision into the destination competitiveness model.

The mechanisms integrating, building, and sharing these values amongst the destination stakeholders in order for them to become reinforced and effectively lead the destination evolution remained unclear. There is an indication of a sustainability approach to driving competitiveness, but the managerial challenges this implies were not discussed and sized. Put simply, there was an important perception of the need to integrate sustainability into the competitiveness process, opening the need for future research.

More recent scholarship on destination management provided important insights suggesting that sustainability is integral and functional to competitiveness. Sustainability was tested as an approach to innovate destination offerings and make destinations exist in highly competitive markets by steering tourists' loyalty through a direct impact on image, quality, and satisfaction (Campòn-Cerro et al. 2017).

In a practice-driven contribution to the destination management debate, Laesser and Beritelli (2013) summarised the viewpoints of an experts' panel warning against sustainable destination development becoming "a catchphrase without deeper meaning" (p. 47). They stated the importance of collectively shaping a sustainable destination vision and stakeholder engagement with long-term sustainability goals. In this piece of work, which is a sort of destination management manifesto, the authors stated:

> Sustainable products and services offer the potential for true differentiation yet the willingness to pay for the attributes that produce sustainable tourism is low. People behave in a sustainable way if the benefits from their behaviours are individually internalised (that is, when they receive benefits from their behaviours), rather than when benefits are externalised (when their behaviours appear only to benefit others) (p. 47).

This pragmatic view on the sustainable destination comes to the core of management, which must deal with shaping and marketing sustainable products and services that (co)create *perceived* value for tourists. According to the authors, this is the basis for building sustainable and competitive destinations: sustainability becomes tangible as intrinsic to products and services, and it contributes greatly to their competitiveness by boosting differentiation. This supports the view of sustainability as a *driver of competitiveness*.

This conceptualisation of the sustainable destination reduces complexity by considering destination fragments, such as sustainable products and services, and the groups of stakeholders working for them. The problem is how to rebuild the fragments in a long-term sustainability vision that engages all stakeholders, including residents and those without any economic and commercial interests in tourism, avoiding *greenwashing*—perceived and/or actual—effects. In addition, there seems to be a predominant role for the firms that are directly in charge of designing destination products and services. Besides the call for tourism stakeholder engagement and participation, there is room to wonder how and to what extent the rest of the destination stakeholders can realistically be involved in sustainable destination development.

3.1.2 Stakeholder Engagement and Sustainable Tourism Governance

In management studies, engagement is the manifestation of behaviours towards a firm, an organisation and its brand that transcends the commercial relationship (Verhoef et al. 2010). A manifestation of behaviour towards a focal object, organisation, or purpose, together with active cognitive processing and emotional bonding, frame stakeholder engagement towards innovation (Loureiro et al. 2020).

Beyond the original focus on customer engagement (Bowden 2009; Verhoef et al. 2010; Brodie et al. 2013), as discussed above, multi-stakeholder engagement largely trickled down into various streams of literature to investigate the engagement of all the different actors who affect or are affected by an organisation's goals and actions (Freeman 1984). The engagement was discussed in relation to the set of practices undertaken by an organisation to positively involve multiple stakeholders in organisational activities based on mutually beneficial schemes of relationship building (Phillips 1997). Scholars' interest in sustainability further increased attention on stakeholder engagement, connecting with the agenda of innovation and value creation (Greenwood 2007; Loureiro et al. 2020). Critical studies have also revealed the limited dialogic exchange in stakeholder engagement for organisations overlooking human agency and the power dynamics constraining effective dialogue (Passetti et al. 2019). Stakeholder engagement may result in consensus building and control processes, having a limited democratising effect on governance (Hernandez 2006).

In line with these evolutions in the academic debate, destination management scholars have increasingly acknowledged the multi-stakeholder nature of destination development based on the critical role of tourism actors' interdependences in boosting destination competitiveness (Dwyer and Kim 2003; Ritchie and Crouch 2003; Beritelli 2011) and sustainable tourism development (Martín et al. 2018; Cheer et al. 2019; Cheung & Li 2019).

Stakeholder engagement was said to be the precondition for tourism networking and collaboration partnerships (Lavandoski et al. 2018), which may impact regional development by boosting connections between tourism and other regional stakeholders (Hall and Mitchell 2000). Stakeholder involvement is the fundamental base for successfully designing and implementing sustainable tourism strategies (Hall 2008). On the one hand, the limited engagement capacity of the tourism system may cause a delay in destination development (Festa et al. 2020); on the other hand, engagement—when assisted by effective technological tools—may trigger stakeholders' empowerment, co-creation, and co-design of innovative experiences (Sigala and Hallen 2019).

A critical agenda on stakeholder engagement for the sustainable destination deserves attention to frame the effective contribution of the engaged stakeholders to the sustainable destination. Engagement, participation, and governance are salient and interconnected faces of the sustainable destination. Nevertheless, how can we ensure that the diverse stakeholders' interests, visions, and resources are integral to decision-making in an effective "dialogic exchange" (Passetti et al. 2019)? How to guarantee that stakeholder engagement does not end up in "politicking" (Ooi 2004) the key decision makers' perspective towards consensus building? How can engagement evolve towards a strong stakeholder sense of ownership of the destination development trajectory? How to reach destination governance arrangements that guarantee "the rights of local communities" (Higgins-Desbiolles et al. 2019) in the face of an increasingly global tourism demand and supply chain? This means that the "organisational capacity for sustainability" is central to sustainable tourism development, and this has to extensively deal with "human resource development", as Rasoolimanesh et al. (2020) stated.

From traditional top-down approaches to managing the sustainable destination, contextual and embedded governance (Go and Trunfio 2011) was discussed in destination management as an alternative to centralised decision-making. Embedded governance relies on a cohesive group of stakeholders engaged with destination competitiveness and sustainability. Regional-global players' connections are a salient part of embedded governance. Global players should be acknowledged as critical actors in sustainable destinations (see, for instance, Gui and Russo 2011, about the necessary connections and missing links between cruise tourism, the port authority, and regional stakeholders).

Emphasising bottom-up governance processes bringing together a dispersed capital of knowledge, visions, and perspectives, the quadruple-helix configuration was discussed as a hypothesis for sustainable destinations (Pasquinelli and Trunfio 2020b). This complements the collaborative framework involving government, industry, and university, proposed by Leydesdorff and Etzkowitz's framework (Leydesdorff and Etzkowitz 1996), with citizens' active participation to create the conditions for innovation in local systems (Carayannis and Campbell 2014). The quadruple-helix involves actors beyond the tourism sectors and the destination's geographical and/or administrative borders, boosting agency and responsibility across residents, tourists, commuters, "temporary residents", tourism firms, and international tourism actors playing a crucial role in sustainability.

Yet "the broader the set of actors, the more challenging the activation of [engagement] mechanisms" (Pasquinelli and Trunfio 2020b, p. 5). This goes in parallel with the rise in scholars' interest in the complexity of tourism governance research to address the multiple links in a destination's networked system and interrelations (Farsari 2021). This bundle of links, also connecting different scales in governance arrangements, informs us about democratic processes by giving insights into the modalities and effectiveness of access to decision-making (Bramwell and Lane 2011).

This is possible if a series of concrete determinants are in place, such as transparency, open and public debates on the agenda to be pursued, identification of stakeholders, possible alternative actions, and indicators to forecast scenarios, choose, and then monitor their consequences (Rasoolimanesh et al. 2020). To complement this governance definition, Ruhanen et al. (2010) identified seven governance dimensions, which were accountability, transparency, involvement, structure, effectiveness, power, and efficiency.

Adaptive governance has been increasingly discussed concerning sustainable tourism. It implies social learning (i.e. interaction, collaboration, and knowledge exchange), learning-by-doing and knowledge-building as conditions that allow governance to function and evolve, (re)composing power dynamics and improving the "organisational capacity for sustainability" (Farsari 2021). Policy learning and learning from policy failures should also be considered governance components, providing the basis for a mutual and interacting exchange leading to learning. Learning was overlooked in sustainable tourism studies, even though it potentially helps to explain the difficulty in effectively triggering sustainable tourism systems (Hall 2011).

Empirical analyses and applied research projects are needed to support the implementation of the quadruple-helix model and outline the potential and pitfalls of tourism governance arrangements in their "procedural and/or substantive aspects" (Novy and Colomb 2019:368). Beyond the critical implementation aspects and despite the limited evidence of governance change impacting the sustainability goals (Dodds and Butler 2019; Novy and Colomb 2019), innovative forms of tourism governance for sustainable destinations need to prioritise the development of an organisational capacity, social and policy learning and to pursue a local agency in tourism development (Novy and Colomb 2019; Cheer et al. 2019).

3.1.2.1 The Objective and the Subjective Dimensions of Sustainability

The multi-stakeholder view of destination development and management introduces two analytical perspectives on the study of the sustainable destination, which are: the subjective and objective dimensions of sustainability, focusing on the implications and impacts of tourism functioning, growth, and crisis on the destination; and the passive and active actors of tourism imbalances. Analyses of sustainable tourism and overtourism literature streams introduced the relevance of objective and subjective sustainability dimensions, as discussed in the broader sustainable development debate. Economists revealed and addressed the issue of objectifying sustainability by producing indicators, composite indexes, and models framing tourism imbalances and problematic contexts objectively in relation to social, economic, and environmental domains (Hueting and Reijnders 1998).

The connections between sustainable development and the wellbeing debates suggest the need to consider both the subjective and objective dimensions of sustainability. Wellbeing was studied concerning its objective, psychological, and subjective dimensions, and the same is needed for sustainability, with an opportunity to identify synergies and tensions (Dolan et al. 2007, cited in Laureti et al. 2018; De Neve and Sachs 2020; Qasim and Grimes 2021). In parallel with the pivotal role of human beings and local communities in defining sustainability and tourism imbalances, the subjective dimension cannot be overlooked. Accordingly, subjective and objective sustainability indicators were said to be necessary (Rasoolimanesh et al. 2020).

The objective dimension of sustainability concerns the sustainability performance of the destination and, intrinsically, the quantification of factual tourism impacts. These can be measured and are an intrinsic consequence of the tourism system's functioning, growth, or crisis (e.g. vulnerability to the pandemic crisis, natural disaster, and terrorism attacks). Objective sustainability indicators describe the state of the destination and support tourism planning and management (Asmelash and Kumar 2019; Rasoolimanesh et al. 2020; Arbolino et al. 2021). These are "measures of economic, physical, social or societal reality unfiltered by perceptions and independent of personal evaluations" (Miller, 2001 cited in Rasoolimanesh et al. 2020, p. 7).

The Sustainable Tourism Index (Punzo et al. 2022) statistically framed the economic, social, and environmental dimensions, empirically demonstrating the possibility to quantify and read their intertwining in tourism sustainability. Based on an extensive review of the literature, the pillars of this index included the economic value generated by the tourism system, tourism enterprises' sustainability performance and behaviour, the region's international appeal, tourism demand, presence of cultural heritage organisations, energy and water consumption, sustainable energy and water management, waste management, quality of the environment, and promotion of natural heritage, security, health, mobility, and gender balance. All these pillars are built on indicators sourcing official statistical data from national and regional databases, based on systematic data collection and census data.

Additional indicators emerged in the overtourism literature, suggesting the need to use data from sources other than official statistical institutions and national economic bodies to analyse the factual effects of tourism (Pasquinelli and Trunfio 2020a). For instance, despite their limitations, the news media, sharing economy platforms (e.g. Airbnb about short-term rentals), and ad hoc surveys should be considered. These may be relevant data for sizing factual effects such as the rising cost of living, residents' displacement, the rise of anti-tourism movements, protests and local resentment (i.e. tourism-phobia), commercial conversion, and overcommercialisation of city centres.

A recent review of the literature on sustainable tourism indicators highlighted a lack of indicators concerning tourism governance, which—for some—is to be considered a pillar of sustainability (Timmermans, 2019 cited in Rasoolimanesh et al. 2020). The difficulty in turning indicators and indices into practices for policymaking calls for research advancements (Blancas et al. 2016; OECD 2016; Punzo et al. 2022). The availability of tourism microdata and the necessary time to be collected, analysed, and used before becoming outdated are important aspects to consider about the objective dimension of sustainability.

Besides the technical aspects of data collection and analysis, the objective dimension is not enough for sustainability to be actionable. Beyond their descriptive role, sustainability indicators support tourism planning and management when there is a criterion to say that the resulting picture is "good enough" to assure a sustainable trajectory or "negative enough" to motivate change (ideally, suggesting what direction and types of actions should be undertaken).

The opportunity for tangible support can be further complicated. Although a set of indicators can be usefully applied to all destinations, the sustainability issues and aspects to be monitored may take very different shapes in different geographical contexts, imposing the use of ad hoc sets of indicators. On the other hand, different indicators may be needed to produce actionable knowledge in the different destinations. This may depend, for instance, on tourism governance in a certain context. Scale (i.e. local, regional, and network) and configuration (e.g. multi-level, formal, institutional, and financial links between the DMO and local/regional governments) influence the usability and power of sustainability indicators (Biermann et al. 2017, cited in Rasoolimanesh et al. 2020).

Certainly, dynamic monitoring comparing a dashboard of sustainability indicators over time can be a valuable support for decision-making. It allows comparing across destinations and analysing their evolutionary trajectory over time. At the same time, this practice can be exerted in long-term frames constraining prompt responses to tourism imbalances, which may be exacerbated over time.

The critical aspect concerning the definition of thresholds for interpreting sustainability indicators and indexes remains a key issue. "How much is too much?" when we speak about tourism and negative tourism impacts (CREST 2018), and in line with this logic, "how little is too little?" in terms of positive tourism impacts for compensating the negative sides of the tourism phenomenon. How can we state that a destination is underperforming in terms of sustainability? Against what criteria can we have a sense of direction in destination management towards more sustainable tourism development?

Defining a quantitative threshold of sustainability is not easy. It can hardly be defined as "objective" as there are judgements to be made that are relative to the destination, to specific contingencies at local and global levels, and to the mindsets of those individuals defining the threshold (McCool and Lime 2001). As we have said, degrees of arbitrariness are unavoidable, and subjectivity ends up determining the objective dimension of sustainability.

In this sense, the objective dimension of sustainability is closely connected with and complemented by the subjective dimension of sustainability. Put simply, the subjective dimension consists of the perceptions of the destination, tourism evolution, tourism's contribution to the local economy, and its impacts on the destination community. Subjective sustainability indicators were recently defined as "usually refer[ring] to attitudes, experiences, perceptions, and the satisfaction of residents and tourists" (Rasoolimanesh et al. 2020, p. 7).

The diverse stakeholders have different perceptions of the destination that, for local stakeholders, also coincides with the place in which they live and work. On the one hand, residents' feelings about how tourism influences their quality of life and, on the other, visitors' perception and satisfaction with their destination experiences are central to sustainable destination development. Although less investigated in the literature, tourism entrepreneurs' perceptions of the destination competitiveness, the attention paid to tourism by local government, and policymakers' perceptions of locals' consensus for tourism initiatives are also important aspects of the subjective dimension of sustainability. There is also room for indicators integrating stakeholder perceptions of tourism governance, which is increasingly considered a critical component of sustainability.

Many studies addressed the subjective dimension of sustainability. One of the first contributions was the Irritation Index (or Irridex) by Doxey in 1975, who modelled residents' emotional reactions to tourism growth, in a sort of "emotional lifecycle" of the destination. The model described the change in residents' attitude to tourism in different stages of the destination lifecycle when the intensification of social, economic, and environmental impacts is perceived (Pavlić and Portolan 2016).

Four stages framed residents' reactions, ranging from euphoria for the growing tourism development opportunities to apathy, irritation, and even antagonism. This is the ultimate phase of residents' irritation due to feeling the loss of their right to live in their place destination. This view couples with more recent scholarship analysing overtourism as a narrated phenomenon by online news media, which spread a narrative of tourism impacts and imbalances, globally echoed and impacting public opinion. Consequently, tourism imbalances are increasingly addressed in similar ways worldwide, regardless of local specificities (Pasquinelli and Trunfio 2020a).

Both tourists' (Yeh et al. 2012; Jin et al. 2016; Rasoolimanesh et al. 2017) and residents' perceptions of overcrowding (Teye et al. 2002) were under scrutiny, revealing contrasting evidence. Negative perceptions of overcrowding (Fleishman et al. 2004; Perdue et al. 1999; Stewart and Cole 2001; Riganti and Nijkamp 2008) were counterbalanced by positive feelings about congestion understood as a sign of a positive place reputation and a significant opportunity for visibility in social media. Overcrowding was also seen as an opportunity to live a memorable experience shared with many other people (Addis and Holbrook 2006; Neuts and Nijkamp 2012), as in the case of international festivals (Kim et al. 2016).

"Tourist ghettoes" were under scrutiny. Residents' positive or neutral attitudes to tourism were witnessed (Dumbrovská 2017). Specific individual attributes, values, and circumstances play a role in shaping perceptions (Dumbrovská 2017; Muler Gonzalez et al. 2018; Martín et al. 2018; Cheung and Li 2019; Gutiérrez-Taño et al. 2019), such as individual economic involvement in the tourism system (Andereck et al. 2005).

Are overcrowded tourism precincts necessarily a signal of an unsustainable path of tourism development? Next to the objective dimension of sustainability, the need to consider the subjective side is evident, to acknowledge the centrality of a local community's perspectives on local sustainable development. How and to what extent do the subjective and objective dimensions guarantee for future generations' interest? The objective dimension can help in not excessively prioritising contemporary stakeholders, their perceptions and feelings, at the expense of future generations' needs. On the other hand, stakeholders' values and beliefs can be the necessary guarantee to go beyond the discussed limitations of any attempt to quantify the tourism imbalances.

The objective and subjective dimensions of sustainable destinations, their synergies and tensions, require further efforts of conceptualisation and measurement. The lack of a consolidated theoretical background holding together these two dimensions creates the risk of "false objectification" or "ephemeral subjectification" of the sustainable destination. Instead, identifying these two dimensions has evident managerial implications, helping to visualise the need to consider the destination from the physical, organisational, and perceptual perspectives.

3.1.2.2 Identification of Passive and Active Stakeholders of the Sustainable Destination

Stakeholder theory has been extensively adopted in business studies to frame sustainability. The company is understood as a set of relationships amongst internal and external actors that affect or are affected by the business's operations and functioning (Freeman 1984). The stakeholder perspective helps to analyse the processes occurring at the intersection of social, business, environmental, governance, and ethical dimensions (Laplume et al. 2008). Diverse stakeholders are central to value creation (Lüdeke-Freund and Dembek 2017). Multi-stakeholder collaborations ensure the opportunity to pool the necessary resources and expertise (Donaldson and Preston 1995; Hörisch et al. 2014) in a collective effort, relying on stakeholders' diverse mindsets, attributes, and resource endowments.

The motivation for collaboration resides not only in shared values but also in mutual benefits for all stakeholders, linked to their appreciation of each stakeholder's action and active involvement (Freudenreich et al. 2020). That is, value is created "with and for" the various stakeholders (Freudenreich et al. 2020). The lack of reciprocal payoffs would produce a disengagement, detrimental to value creation and constraining sustainability.

These mechanisms, framing value creation in business models, explain the importance of a multi-stakeholder perspective on sustainability because of the centrality of the different roles and outcomes framing the processes in place in a sustainability framework. On the same premise, the stakeholder perspective is relevant to sustainable destination management. Stakeholders, as has been said, are all those actors that affect or are affected by the local tourism system. We can reasonably speak of a sustainable destination when value is created *with and for* all stakeholders. All stakeholders contribute to the sustainable destination with, for instance, time, investment, participation, cognitive resources, and networks, and gain some value from it. If this does not apply, tourism disequilibria and imbalances will likely emerge.

Table 3.1 presents an ideal state of sustainability through a non-exhaustive hypothesis of value creation *with and for* the different stakeholders. Balance and inclusion are keywords to explain the equilibrium amongst the "weights" characterising the items in the "value creation with" and the "value creation for" columns, across the different rows (i.e. the stakeholders) in the table. Such

Table 3.1 Stakeholder-based view of the sustainable destination

Stakeholder	Value creation *"with"*	Value creation *"for"*
Residents	Common goods (e.g. public space, environment, traditions, and local culture): Reduced use/access Participation ("situated knowledge") Consensus	Cultural exchange Income (second home rentals) Place image and property market value Place image and sense of pride Cultural offering and events Protection and enhancement of local cultural and natural assets Future generations' needs and interests
Tourists/visitors	Experience co-creation (e.g. creative tourism) Responsible behaviour and sense of limit	Cultural exchange Satisfaction Learning Self-fulfilment Future generations' needs and interests
Tourism workers	Expertise Know-how	Job and income Professional growth Job satisfaction Future generations' needs and interests
Government	Investments Public services Planning	Tax collection Job creation Private investment attraction Place brand building Consensus Future generations' needs and interests
Local tourism and tourism-related operators	Expertise and know-how Creativity and business ideas Investments and entrepreneurial risk Local and national taxes Job creation	Income and profits Professional satisfaction Future generations' needs and interests
Global tourism and tourism-related operators	Expertise and know-how Creativity and business ideas Market access and "market mavens" (global connections) Job creation Local and national taxes	Income and profit Future generations' needs and interests
Local non-related tourism operators and organisation	Limiting their use and exploitation of local resources (devoted to the tourism system)	Place image and talent attractiveness (visibility, cultural and natural assets, reputation, etc.) Future generations' needs and interests

equilibrium is the necessary condition for social equity, environmental justice, and the economic empowerment of the destination community.

Remarkably, future generations' needs and interests are integral to this table and present in all the table's rows. The concern for future generations emerges in the table in two ways. Firstly, such concern is a shared value that motivates multi-stakeholder collaboration and guides their behaviours (Freudenreich et al. 2020). Secondly, it is a payoff for stakeholders as a form of reassurance, a sense of commitment, trust, and a sense of belonging when this shared value is met and respected by the other stakeholders.

Disequilibria emerge in two main scenarios. First, when value creation occurs "with and for" only a subset of stakeholders. This means that only a part of the stakeholders (those actors who affect or are affected by the tourism system's operations) contribute to the tourism system (column "value creation with"), whilst the others only benefit from it (column "value creation for"). Second, disequilibria emerge in the case of imbalances between the resources that one or more stakeholders devote to the tourism system and the benefits these gain from it. For instance, residents may refrain from visiting local heritage or natural sites for the presence of tourists whilst not receiving any significant benefit from tourism (in this regard, the overtourism definition based on locals' understanding of tourism-related gains as limited and outweighed by the costs of tourism growth).

This table is helpful for two reasons. First, it suggests the multiple possible configurations of the sustainable destination, depending on the contextual balance amongst the rows and columns. Second, it helps to draw attention to the individual stakeholders and to the fact that balances and imbalances need to be studied (and built, from a practice perspective) by assuming a stakeholder perspective. Many realistic configurations of non-sustainable scenarios emerge from the table. Let us make some additional examples of tourism disequilibria that have been extensively discussed in the literature. Residents might be displaced from tourist areas because of increasing property and living costs, experiencing negative feelings and a sense of irritation because of tourists' exploitation. Global actors (i.e. international intermediaries, OTAs, sharing economy platforms, cruise liners, and the film industry) might exploit local resources without creating numerous and qualified jobs and/or being eligible for significant local or national tax payments. Global stakeholders are "outside the control of destinations and policymakers" (Peeters et al. 2018, p. 18), and ways to include them in sustainable value creation must be found. Tourism workers may be in precarious, poorly paid, or even illegal work conditions.

The concrete challenge of sizing resources and benefits related to each stakeholder further complicates the reading of this table. This connects with the difficulties of quantifying and measuring the objective dimension of the sustainable destination discussed in the previous section. This also combines with the subjective dimension of the sustainable destination (see the previous section) in relation to the key role of the "appreciation" of each stakeholder contribution, motivating stakeholders' collaboration and mutual relationships (Freudenreich et al. 2020).

Appreciation is the word used by Freudenreich et al. (2020) and means a full understanding and recognition that something is valuable and important.

Accordingly, appreciation implies perception, information, and knowledge about the contribution by and payoff for each stakeholder in the tourism system. As has been said, context variables and individual norms and values influence subjective perceptions (Shi et al. 2017; Dumbrovská 2017; Oklevik et al. 2019). *Tools and practices* that circulate data and information, allowing stakeholders to perceive, know, and hence, appreciate these dynamics are at the core of sustainable destination management.

In this regard, scholars have focused on the narrative representation of tourism and tourism impacts and their influence on subjective perceptions. Tourism narratives influence any destination stakeholder's thinking and are likely to impact the degree of *appreciation* of any other stakeholder's contribution to and payoffs from destination value creation. Tourism narratives originate inside the destination, concerning the specific traits of the local tourism phenomenon, or globally, regarding tourism models and impacts that transcend specific contexts. The media do contribute to structuring such narratives. For instance, trans-urban and trans-national grassroots platforms, which usually target job precariousness, inequality, and social exclusion, have started catalysing their attention on tourism imbalances and the negative tourism impacts in urban contexts (Milano 2017).

Tourism studies have focused on the role of the media and news media in building consensus and setting the policy agenda about tourism since framing public opinion and the understanding of the tourism system's role in local development (Hall 2003; Schweinsberg et al. 2017). Furthermore, the media significantly influence the destination image (Gartner 1994). Media focus follows the "issue-attention cycle" trajectory, from hype to decline (Hall 2003), capturing the public debate on a given issue in the initial phase of close attention.

Representation of the overtourism phenomenon made by the international news media (Pasquinelli and Trunfio 2020a) introduced and circulated the word "overtourism" and contributed to place this word in the vocabulary of policymakers and practitioners (Milano 2017). For some scholars, the media shine a negative light on tourism, which was identified as the main actor of neighbourhood change, amplifying and even "exaggerating" fears against tourism and the related gentrification phenomenon (Füller and Michel 2014, p. 1313). The news media proposed cause-effect mechanisms explaining a phenomenon (Fulton 2005). The point is that what is newsworthy is subjectively defined, and influenced by ideologies, cultural beliefs (Dunn 2005), and different economic interests (Fulton 2005).

Analysis of international online news media archives identified passive and active characters in the representation of overtourism (Pasquinelli and Trunfio 2020a). Specific actors are generally identified as the cause of overtourism, such as low-cost carriers, apartment rentals, tourists from booming market segments such as middle-class tourists from emerging economies and millennials, tour operators and cruise companies, and governments (with their inactivity).

Residents are represented as passive actors, the most negatively impacted by the narrated overtourism, with no active contribution to tourism imbalances from their side. The news media's representation of overtourism mostly focused attention on the local origin of overtourism where there are key responsible agents like governments (much emphasis was on government) and tourism businesses, whilst residents emerged as tourism victims. Recent scholarship has analysed residents in overtouristified destinations, demonstrating the variety of roles they may play. The "victim" archetype was described as the type of resident particularly vulnerable to the rise in living costs (Martín Martín et al. 2018), like young and low-income residents in the rental market. On the other hand, the "peaceful activists", "vandals", and "the resilient" were described (Séraphin et al. 2020).

Representations missing nuances in explaining overtourism causes and effects fail to consider the broader societal and urban phenomena that go in parallel with overtourism and boost tourism growth (Koens et al. 2018). Below the surface are issues such as congestion, overcrowding, and gentrification, which may seem exclusively caused by tourism but are deeply connected to territorial and societal developments. Residents may be motivated to move out of the city centre for various reasons: to offer short-term apartment rentals to tourists; to change their lifestyle and consumption models in a changing society (especially after the COVID-19 pandemic). Flexible work arrangements, mobility facilitating and commuting, online shopping are all factors actively pushing forward and accelerating tourism conversion of places. The pandemic may have furthered some trends, creating a continuity between pre- and post-pandemic tourism (Pasquinelli and Trunfio 2022).

To conclude, the identification of sustainable destination stakeholders is based on identifying the roles played by those actors that affect or are affected by tourism system operations. Sustainable destination configurations imply that value is created "with and for" all stakeholders. Otherwise, imbalances emerge. Contributions to and payoffs from the sustainable destination value creation processes need to be "weighted" considering each individual stakeholder perspective, despite the objective difficulty in measuring and attributing such "weights".

Besides the concrete contributions and payoffs, the perceived and appreciated contributions and payoffs matter greatly to the sustainable destination and are influenced by different variables characterising the context, the stakeholders' mindsets, and the role of actors, such as the media and tourism media narratives shaping people's minds. How can we support information and data circulation to make stakeholders develop their perceptions and knowledge? How to create the interpretative capacity for stakeholders? Once the destination system has evolved in the direction of no mutual appreciation amongst stakeholders, preventing any motivation to collaborate, what tools and practices may help to break such vicious cycles?

3.1.3 Sustainable Tourism and Sustainable Destination Management: Key Propositions

To summarise the previous sections, a set of propositions frames the achievements of sustainable tourism studies, drawing attention to sustainable destination management. These propositions help to outline the managerial challenges that sustainable destinations must address. They can be understood as the backbones of the conceptual advancements this book aims to develop, grounding them in the sustainable tourism debate.

The *first proposition* concerns the centrality of destination management for sustainable tourism. As previously stated, there is no sustainability without a managerial effort (decision-making and action taking) that, throughout the destination lifecycles, aims to orient the tourism system towards sustainability goals, integrating and in synergy with the whole local system. The *second proposition* concerns the need to consider sustainability and competitiveness as Janus-faced concepts. Integrative conceptualisations of destination competitiveness are needed, going beyond tourism market stakeholders' views and focusing on destination innovation and how sustainability can drive it.

The *third proposition* is that destination management and sustainable destination management are not a "hard science", not even a pre-given set of golden rules and recipes to learn and apply. We learned about the significant degree of *subjectivity, arbitrariness, and imagination* or *vision capacity* (the capacity to frame the desired state of equilibrium) characterising the concepts guiding the theory and practice of sustainable tourism management (e.g. carrying capacity, acceptable change model, individual tourism perceptions depending on personal and socio-economic characteristics). Any destination management perspective on sustainable tourism implies a constant process of trial and error, compromising choices, reflections on the gaps between image and reality, and constant adjustments and flexibility facing changes in the region, society, and the global tourism system.

Connected to this, *the fourth proposition* concerns destination stakeholders as a pivot of sustainable destination management: multiple and extremely diverse stakeholders, internal (i.e. local operators and residents) and external (i.e. international operators and visitors) stakeholders, tourism and non-tourism actors that *affect or are affected* (Freeman 1984) by the tourism phenomenon and by the evolution of the destination. Last, we cannot forget that sustainability means "meeting the needs of the present without compromising the ability of future generations to meet their own needs" (United Nations Brundtland Commission, 1987). Hence, the sustainability concept includes another macro-category of stakeholders, which is future generations, who will be affected by the tourism phenomenon, its manifestations, and durable impacts, and who will affect tourism systems with their resources (e.g. human capital).

The *fifth proposition* concerns the engagement of all these possible stakeholders in building the frame that guides and constantly inspires sustainable destination management. Individual responsible behaviours present limitations in pursuing systemic tourism change oriented to sustainability. At the same time, we must consider how challenging it is to achieve functioning stakeholder participation and involvement. The *sixth proposition* (deriving from the fifth) regards the central role of sustainable tourism governance that organises and orchestrates stakeholder participation and involvement. Governance configurations (Trunfio 2008) remain an open and significant field for research. "Responsibility, control, governance" (Cheer et al. 2021) are the bases for sustainable destinations: these three words summarise the central role of individual responsibility, choices, and behaviours, but also the need to boost collective processes and sustainable tourism governance, steering, orchestrating, monitoring, and controlling the destination's evolutionary path.

3.2 Recent Advancements in the Sustainable Tourism Debate: The Overtourism Debate

A new wave of interest in sustainability in tourism studies emerged over the last few years around the notion of overtourism which, before the pandemic outbreak, became central in tourism debates in the media, amongst policymakers and academics. From a theoretical perspective, the overtourism debate was positioned in the sustainable tourism agenda (Pasquinelli and Trunfio 2020b). For some, it was "old wine in new bottles" (Dredge 2017), echoing the previously mentioned study by Rosenow and Pulsipher, who in the 1970s framed "visitor overkill" for the first time.

Recent overtourism scholarship has reconsidered the theoretical frameworks inherited from the past and from the longstanding debate on tourism impacts to reframe tourism development models towards overcoming tourism conflicts and imbalances (Milano 2017; Koens et al. 2018; Muler Gonzalez et al. 2018; Pasquinelli and Trunfio 2020b; Diaz-Parra and Jover 2021). Compared to the general debate on sustainable tourism, the overtourism debate has exclusively focused on the downsides of tourism intensification. It limits its viewpoint to framing and addressing tourism issues and acknowledges the impossibility of thinking of tourism without thinking of the negative impacts it produces.

The overtourism debate can be summarised according to the following streams of research:

(a) The articulation of the overtourism phenomenon in different geographical contexts, focusing on its prominence in urban contexts but, at the same time, its relevance in non-urban destinations (Peeters et al. 2018; Milano et al. 2019; Séraphin et al. 2020; Pasquinelli and Trunfio 2020a).
(b) The overtourism capacity to reshape the economic, physical, and social landscapes of cities and regions (Peeters et al. 2018; Koens et al. 2018; Séraphin et al. 2020; Pasquinelli and Trunfio 2020a).
(c) The emergence of tourism-phobia, anti-tourism, and tourism saturation in various contexts (Colomb and Novy 2016; Milano 2017; Koens et al. 2018; Martín et al. 2018; Séraphin et al. 2018).
(d) Serious tourism impacts on residents' quality of life and deterioration effects on the value of the tourist experience (Cheer et al. 2019; Cheung and Li 2019; Dodds and Butler 2019; Novy and Colomb 2019).

These were the crucial aspects that, pertaining to the broader and longstanding sustainability agenda, were reframed and reconsidered by the overtourism debate considering contemporary societal, economic, and environmental scenarios. The local community perspective is predominant in all the available definitions of overtourism (Colomb and Novy 2016; Milano 2017; Séraphin et al. 2018; Milano et al. 2019). The local community is the object of observation to identify and study overtourism, focusing on the permanent changes to local life and to people's sense of place, weakening socio-cultural connectivity, and the (over)exploitation of local environmental and cultural resources.

Impactful forms of tourism pushing overtouristification processes are also defined based on locals' understanding of tourism-related gains as being limited and outweighed by the costs of tourism growth (Cheer et al. 2019). The limit of this definition is the lack of capacity to capture the many different identities, interest groups, and individual viewpoints comprising the local community. This concerns a crucial point in the sustainability agenda regarding the balance of costs and gains from tourism but, most especially, their allocation amongst all the diverse stakeholders of the local community.

Beyond understanding overtourism determinants and impacts (Koens et al. 2018; Oklevik et al. 2019; Gravari-Barbas and Guinand 2017; Kim and Kim 2020; Szromek et al. 2020), destination managers, policymakers, and scholars should be concerned with policy approaches, managerial practices, and tools to confront overtourism (despite the difficulties to define and size the phenomenon) and the "overtourism hazard". These are of interest to both mature and emerging destinations (Pasquinelli and Trunfio 2020c).

Cities like Barcelona, Berlin, Venice, and Amsterdam are icons of the phenomenon and the social and political fight against it (Milano 2017; Minoia 2017; Fava and Rubio 2017). These cities started reflecting on how to manage overtourism and counteract the negative impacts of its rapid growth. There is a learning opportunity for those cities and towns which, undertaking (or willing to undertake) a tourism development path, can anticipate tourism impacts and negative effects. As previously stated, overtourism concerns not only urban centres and the most reputable and appealing destination brands worldwide but also second-tier cities and smaller towns, rural, mountain, and coastal destinations (Peeters et al. 2018; Milano et al. 2019; Séraphin et al. 2020).

Although the geographical context matters greatly to development models (analysis and design) and their implications, overtourism mechanisms may be recurrent and similar in different contexts, although taking place at different scales and timings. Examples of mechanisms that may occur in different geographical contexts are the overcrowding of public spaces, tourists' pervasiveness and inappropriate visitor behaviours, the physical touristification of old towns and other highly visited areas, residents' displacement from residential areas caused by increasing living costs, and pressure on local environmental resources (Koens et al. 2018; Dodds and Butler 2019).

The analysis of overtourism's causes and effects presents a significant degree of development, capitalising, and furthering of the sustainable tourism agenda. Nevertheless, the available research on the *managerial approaches and tools* to overcome overtourism imbalances and address sustainable development remains a less developed field. Although the literature on its causes and effects on destinations is consolidated, there has been a limited and less robust scholarly engagement with the "so what?" question. What to do to prevent the overtourism phenomenon? What conditions allow destinations (and their stakeholders) to make such decisions about tourism evolution and to act? What actions are taken by destination managers and policymakers to deal with overtourism and pursue sustainability? What are the actions' effects and implications? Furthermore, what is their impact on the different destination stakeholders? These questions remained largely untapped in the existing literature.

One answer provided by scholars concerned the necessary integration of tourism policies with the local sustainable development framework, making tourism intertwined and functional with the other sectors, socio-economic and cultural activities in the region (De Jong et al. 2015; Bellini et al. 2017). In other cases, a necessary step towards a radical rethinking of tourism development models was stated. This included considering the tourism de-growth hypothesis. Scarcely achieving local consensus, especially amongst tourism entrepreneurs and the actors who benefit from tourism-related economies, such a hypothesis concerns the need to reduce incoming tourist flows, prioritising equity and inclusion over the market imperative (Higgins-Desbiolles et al. 2019; Milano et al. 2019). "Ambidextrous management" was, then, postulating the opportunity to address overtourism by exploiting the derived negative situations as moments for exploring and creating positive conditions for shaping innovative tourism models (Séraphin et al. 2020).

Single actions to address overtourism were discussed in the literature, although these adopted a narrow analytical focus. Optimisation strategies were discussed, aimed at stabilising tourist numbers whilst making efforts to increase income opportunities (Oklevik et al. 2019). Increasingly, attention was drawn to visitor management (Kebete and Wondirad 2019) and the need to impose quantitative limits on visitors to a destination or certain sites within the destination (Muler Gonzalez et al. 2018; Séraphin et al. 2018). In this regard, dispersal strategies were under scrutiny, arguing the need to design new tourism products to alleviate pressure on mass tourism sites and peak seasons (Peeters et al. 2018). This, however, should be carefully considered in light of the potentially negative effects of drawing tourists' attention to unbeaten tracks that may not be resilient to tourist penetration (Füller and Michel 2014). Destination rebranding was also discussed, attempting to enlarge visitors' mindsets and expectations about the destination through brand expansions (Séraphin et al. 2019). Transport, mobility, and infrastructure investments were said to be necessary to reduce visitor pressure (Koens et al. 2018).

To reduce residents' overtourism perceptions and resident-visitor conflicts, various solutions were proposed, such as monetary compensation (as a fiscal solution), incentives for local employment, and community engagement with tourism, such as community festivals, as cultural exchange opportunities (Postma and Schmuecker 2017; Peeters et al. 2018; Séraphin et al. 2019). Beyond the single solutions and actions, there is a need to build broader theoretical frameworks helping to interpret the nature and implications of these actions, transforming them into levers of destinations' structural change.

Destination management scholars must provide a concrete contribution in framing interpretative and evaluative schemes to better understand the different possible actions. This would guide the adjustment of existing tourism models and destinations' transition towards sustainable tourism models (Koens et al. 2019). Distinctive characters, implications, limitations, and potential side effects of these actions have to be assessed in the broader frame of tourism development models. This remains a crucial and largely unexplored side of the sustainability agenda in tourism studies.

Conservative and radical approaches to addressing overtourism were discussed in the literature (Pasquinelli and Trunfio 2020a), an interpretative effort to identify broader frames for explaining action-taking. The conservative framework is set on the tourism-led growth hypothesis (Brida et al. 2014, cited in Pasquinelli and Trunfio 2022). The actions that belong to this approach do not challenge tourism growth as an overall goal and preserve the consolidated tourism models in the destination, so maintaining "business as usual" (Dwyer 2018). The conservative framework is translated in specific terms into technical mitigations of the negative impacts of tourism growth. Visitor expenditure and on-site consumption need to be assisted by sufficient tourist inflow size, which makes tourism and its contribution to the local economy vital.

In contrast, the radical framework postulates that a transformation of the tourism system is the only viable approach to addressing tourism disequilibria (Milano et al. 2019). The most practice-oriented indications concern the need to shape new forms of the local community's engagement in local development (Higgins-Desbiolles et al. 2019) and new forms of tourism governance (Novy and Colomb 2019), privileging inclusivity and the pursuit of local stakeholder wellbeing (Joppe 2018). The radical rethinking of tourism development implies a "transformative tourism" in line with principles of equity and inclusion (Milano et al. 2019).

3.3 Addressing Tourism Imbalances: Normative Enforcement, Visitor- and Stakeholder-Oriented Destination Management

The existing literature identified regulation, management, and marketing tools (Pasquinelli and Trunfio 2020a; Pasquinelli and Trunfio 2022) to describe the different approaches embodied by single actions. According to the original categorisation (Pasquinelli and Trunfio 2020b), *regulatory measures* concerned the deployment of normative power to impose rules like tourist taxes, access limitation, licensing for retail, restaurants, and hospitality activities. The second category was *management,* which implies "doing for orienting" the destination offering and tourists' demand towards sustainability without exercising a coercive power. It concerns "the building of tourism 'products' anticipating the market, by involving visitors and residents in co-creating the destination" (p. 3). The third category was *marketing*. It was defined as "the analysis of the market informing the construction of tourism products and the efforts of communicating, sharing, and delivering their value" (p. 4).

Table 3.2 presents a revised version of this categorisation which expands the identified approaches as follows: *normative enforcement, visitor-oriented destination management, stakeholder-oriented destination management.* The reframing of the original categories was meant to advance insights into the distinctive characteristics, limitations, and potential issues of the different approaches to addressing tourism imbalances within a destination. They represent distinct, yet possibly coexisting, approaches to sustainable destination development, and connect the findings from the academic overtourism inquiry with the broader sustainability research agenda in tourism studies.

3.3.1 Normative Enforcement

Normative enforcement confirms the role of regulation as a possible way to address tourism imbalances in the destination. It is based on the possibility to effectively control respect of the rules and effectively sanctioning those not respecting the limits and, generally, the norms that public authorities define and can impose by enforcing their institutional and administrative powers. The power to define access limits and, on the other hand, the capacity to enforce norms are key issues. The point that this first category wants to make is the importance of stating rules and the centrality of the effective opportunity to make the rule respected. In this regard, thinking not only of the rules but also of the tools and conditions to create to make the rules respected is crucial. With regard to the definition of rules, arbitrariness may be a limitation when revealing the subjective and even ideological nature of some restrictions, as in the case of restrictions to the opening of "non-authentic" commercial activities to preserve the prestige of city centres (e.g. McDonald's in the historical town of Florence or "Nutella shops" in Amsterdam).

As previously discussed in this chapter, the consolidated literature on sustainable tourism echoes the controversial definition of access limits based on the longstanding debate on carrying capacity (Lindberg et al. 1997; McCool and Lime 2001), estimation of the visitation optimum (Pásková et al. 2021), the acceptable change model (McCool 1994), and similar planning and managerial tools (McCool and Lime 2001). All these conceptualisations are based on the identification of the tolerable destination load. What is "tolerable", though, depends on many factors, including personal interests, individual perceptions, values, and time (e.g. tolerability thresholds may change daily, depending on times when there is overcrowding, and in different seasons and over the years, depending on many societal and environmental factors). Rules and norms definition is also linked to the political continuity (or discontinuity) of local and regional governments. This aspect may make these rules temporary experiments, leaving limited effects on the destination development path, especially if it creates significant local discontent and

Table 3.2 Intervention approaches to the sustainable destination development

Approaches to the sustainable destination: three categories	Actions	Distinctive character	Limitations and potential issues
Normative enforcement	Taxes (e.g. tourist taxes) and monetary compensation to residents (fiscal lever) Financial incentives and rescue packages for independent commercial activities Licensing (hospitality, retailing) Normative limitation to short-term rentals and sharing economy Banning "non-authentic" commercial activities such as fast food and souvenir shops Limited access to areas and tourist sites Incentives to local employment	Control Sanctions Political-administrative power, central role of public institutions, and political elite	Effective capacity to control and sanction, effectiveness in defining limits, consensus, the politics of limit definition, political influence, and instability
Visitor-oriented destination management	• Market intelligence for product design. Educational programmes for responsible travellers Preference building programmes Promotion & communication	Market-oriented Activate and reactivate demand	Bias on visitors as central actors of the sustainable destination
Stakeholder-oriented destination management	New alternative product and experience design for visitor dispersal (anticipating the market) Building Centre-periphery relationships Training programme (for tourism and tourism-related operators) Optimisation strategies Investment in infrastructures, transports, and mobility Community festivals and other community engagement initiatives Rebranding and destination brand expansion	Stakeholder-oriented, destination community-oriented Innovation of products, processes, and organisation Stakeholder engagement, co-learning, and co-creation Boosting stakeholder behaviours and mindset change	Identification of the "credible orchestrator" Capacity of the orchestrator to catalyse interest and engagement/creating value for the different stakeholders Capacity to make stakeholders share values, approaches, and ideas Identifying modalities to make stakeholders gain and perceive (and value) the sharing practices payoffs (also in a crisis when the costs of collaboration may be perceived as higher)

Source: Séraphin et al. 2019; Koens et al. 2018; Peeters et al. 2018; Muler Gonzalez et al. 2018; Dodds and Butler 2019; Oklevik et al. 2019; Kebete and Wondirad 2019; Séraphin et al. 2019; Pasquinelli and Trunfio 2020b

the destination community does not understand the importance and validity of these limits and restrictions.

3.3.2 Visitor-Oriented Destination Management

Visitor-oriented destination management consists of market-oriented management of the destination according to which market intelligence and the focus on the visitor (his/her needs, preferences, and expectations) are prioritised in designing tourism products. Tourists (potential tourists and destination visitor streams, Andersson 2017) are the starting point for developing destination products. By virtue of the value they (co)create with the customer, these products might be capable of shaping tourists' preferences and driving visitors to adopt responsible behaviours. The value perceived, expected, and experienced by the tourist (once on-site) is the precursor and driver of responsible behaviours, contributing to destination sustainability.

Promotion, communication, and educational programmes concerning the sustainable destination product or the sustainability content of destination products, inform visitors, create interest, and boost a positive travel attitude. These actions configure a destination marketing approach aimed to shape valuable and sustainable products for visitors. In line with responsible tourism, the individual tourist selects and behaves responsibly, contributing to the sustainable destination (Goodwin 2019).

The strategic and operative marketing challenge is twofold. On the one hand, it consists in identifying and dealing with tourists who actively look for sustainable travel options, are motivated by these (e.g. ethical seekers, see the Visit England toolkit or the Failte Ireland National Tourism Development Authority green marketing toolkit[1]), and feel a sense of limit (Franch et al. 2008). We can speak of tourist-driven sustainability in this case. The challenge is to "harness the consumer's pro-environmental values and social preferences" and position products in the market by imposing a premium price (Wymer and Polonsky 2015, p. 7). On the other hand, the marketing challenge consists in dealing with and responding to a mass tourism market for which sustainability is not central and not even in focus. "Feel good" consumers prioritise unique and memorable experiences and are open to the sustainability content and aspects of their vacation and visit whenever these positively contribute to their experiences. In this case, we need to understand how to assist customer decision-making in making their preferences for green and sustainable products emerge and strengthen (supply-driven sustainability, Wymer and Polonsky 2015).

Accordingly, the distinctive character of this market-driven managerial approach is the aim to activate responsible tourists' demand, a well-defined objective to pursue

[1]Available at: https://icrtourism.com.au/wp-content/uploads/2012/09/Visit-England-tool-kit.pdf; https://www.failteireland.ie/FailteIreland/media/WebsiteStructure/Documents/2_Develop_Your_Business/1_StartGrow_Your_Business/Green-marketing-toolkit.pdf. Last access 1 January 2023.

local tourism players, beyond the abovementioned challenges. The evident limitation is the bias in favour of the customer (the tourist) as if the tourism imbalances could be addressed and solved by responsible tourists' individual choices. This approach implies the capacity of individual tourism businesses (in hospitality, mobility, cultural sectors etc.) to frame sustainability and reach an understanding of what sustainability means in the local context. Besides, the implication is that tourism businesses have the capacity (cognitive and financial resources) to put sustainability into practice in the shape of valuable tourism products. The critique of constraining sustainability to the "feel-good" practices of individuals (both service providers and customers) applies, with the consequent need to broaden our views on the sustainable destination. This is viable by reaching robust insights into the local-global tourism system (Burrai et al. 2019; Cheer et al. 2021) and the multiple, internal and external, private and public stakeholders (see above, Frey and George 2010).

3.3.3 Stakeholder-Oriented Destination Management

In continuity with the management approach introduced in the overtourism literature (Pasquinelli and Trunfio 2020b), *stakeholder-oriented destination management* complements the visitor-oriented destination management approach. It stresses the importance of relying on the stakeholder engagement (visitors, residents, tourism businesses, public authorities) in boosting a sustainability-oriented collective mindset and collaborative action. Instead of coercive power, engagement and sharing are the triggers for this managerial approach to the sustainable destination. As listed in Table 3.2, the distinctive character of this approach is the focus on the destination community that innovates products, processes, and organisational relationships in the tourism system in pursuit of sustainable destination development.

Examples of actions pertaining to this managerial approach are alternative and innovative product design for tourism dispersal, envisioning the connections between centres and peripheries, cities and regions, crowded and less known areas, investment in infrastructure, transport, mobility, facilitation of dialogue and exchange practices involving the different stakeholders (including visitors-residents), and training programmes and learning opportunities for private and public actors.

The distinctive character of this approach is engaging the destination community in co-learning (Colazzo et al. 2008) and co-creating the sustainable destination. The rationale is acknowledging the fact that the destination is an ever-evolving system composed of multiple evolving entities, such as resources and stakeholders (organisations and human beings) who learn and may achieve a shared vision of local tourism development.

The critical side of this managerial approach is the difficulty in triggering and orchestrating the stakeholder engagement process in an effective and steady manner over time. Capturing stakeholders' interest, demonstrating the value and payoffs of stakeholder engagement and collaborative practices for the sustainable destination, sourcing knowledge and information inputs from the different stakeholders, and considering their perspectives and destination perceptions (e.g. quality of life is highly subjective), are not easy tasks. They require the emergence of a capable and credible orchestrator, as suggested by the extensive debate on tourism governance (see Trunfio 2008), which plays an essential role in the sustainable destination.

Stakeholder-oriented destination management is deeply intertwined with sustainability marketing. Font and McCabe (2017) stated that, although sustainability concerns in tourism marketing are less developed than in the overall marketing literature, the emerging sustainability marketing in tourism is:

> the application of marketing functions, processes and techniques to a destination, resource or offering, which serves the needs of the visitor and stakeholder community today and ensures the opportunities of future visitors and stakeholders to meet their needs in the future (p. 871).

Two sustainability marketing approaches were framed: market development and product development. *Market development* works on the concept of sustainability-driven consumerism, putting into the market products with a sustainability content that is visible to consumers. Diverse degrees of centrality in the value proposition characterise products' sustainability attributes, depending on the consumer target (e.g. ethical seekers or "feel good" consumers). In both cases, however, the point is to propose products that, capable of creating value for the consumer, are visibly promoted as being sustainable to increasingly encourage tourists towards responsible behaviours (the links with visitor-oriented destination management are evident, see above).

In contrast, the *product development* approach to sustainability marketing resonates with stakeholder-oriented destination management (Table 3.2). It focuses on the design, development, and constant improvement of a destination's sustainable products and experiences to achieve a more sustainable destination model and normalise sustainable tourism consumption. Whilst communication highlights the personal benefits for the visitor, sustainability is not necessarily communicated and marketed, since it is conceived more as a prerequisite for product existence than a component of the value proposition to be marketed. This is a crucial aspect allowing the destination community to work on and invest in sustainability, considering a long-term horizon, engaging with design, and constant improvements to the destination offering. This approach reduces the risk of greenwashing effects for the destination and its stakeholders and boosts collective thinking and public debate towards the emergence of a locally tailored sustainability vision. This approach acknowledges that such a process necessarily takes much time and effort to coalesce.

As shown in Table 3.2, destination branding, rebranding, and brand expansion are part of the stakeholder-oriented destination management approach. In line with a stakeholder-based view, place brand building was defined as an open, social, and multidirectional process of meaning creation (Govers and Go 2009; Kavaratzis and Hatch 2021). Place branding literature substantially agreed on the role of stakeholder involvement and participation in avoiding conflicts over place representation (Klijn et al. 2012; Eshuis et al. 2014) and counter-branding reactions (Pasquinelli and

Maiello 2015). Place brand building was conceptualised as an orchestrated yet widely participatory process, as much as possible distant from the technocratic and top-down selection of brand attributes and values. As a result, the destination brand should be ever-evolving, dynamically emerging from iterative interactions amongst the stakeholders (Kavaratzis and Hatch 2021).

Echoing the difficulties of general stakeholder-oriented destination management, rebranding, and brand expansion integrating stakeholders' visions for the sustainable destination are extremely challenging. There is a need for further research and advancements in the methods, techniques, and tools to make effective stakeholder engagement occur. This was recently investigated in relation to internal marketing steering residents' role of "critical collaborators" in destination governance towards sustainability (Zhang et al. 2022).

The destination brand may play a fundamental role in the destination community, helping stakeholders visualise the development process, and identify with its aims (Pasquinelli 2014). In contexts of discontinuity and crisis, place branding may even result in brand experimentation (see below) which, though short-term, may give voice to alternative views on local tourism development and trigger the involvement of stakeholders who have remained marginal to destination evolution (Pasquinelli et al. 2022). Research is needed to further explore the hypothesis of place brand experimentations assisting sustainability-oriented innovation processes and the tools, conditions, and modalities that allow this.

3.4 Integrating Smartness and Sustainability: The Missing Links

This section reconnects the sustainable tourism debate with the smart destination literature to discuss technology-driven innovation and its contribution to envisioning and implementing a sustainable destination. First, the academic perspective and, second, the policy domain perspective will be discussed.

3.4.1 The Academic Debate

The literature addressed the link between sustainability and smartness in two main ways. First, sustainability is implicitly *embedded* in the smart tourism destination concept. As Gretzel put it:

> Smart tourism describes a form of tourism development that takes advantage of advanced technologies (notably sensors, wireless communication networks and big data analytics) to achieve sustainable development goals (2022: p. 3).

Smart tourism and the smart destination provide a framework for investigating tourism innovation at a territorial level, highlighting the role and contribution of technologies to tourism development. Beneath innovation, the role of multiple actors, participation, engagement, and social capital are necessary aspects. These aspects, as discussed above, are also at the core of the sustainable destination. However, the smart tourism and smart destination agenda did not always go further than mentioning sustainability as a taken-for-granted implication and there are limited contributions that deeply and critically engage with the concept of sustainability in relation to smart destinations. Besides, although sustainability has been increasingly mentioned in the smart tourism debate, the link with sustainability aims is not crystallised into a clearcut definition of smart destination which is, in fact, still missing (Shafiee et al. 2019; Gretzel 2022).

The smart destination relies on technologies, people, and institutions to create public value (Desdemoustier et al. 2019), which echoes the stakeholder perspective on value creation and sustainability previously discussed in this chapter. Besides technologies, the smart destination implies significant investments in human and social capital, a condition for sustainable economic growth and quality of life (Caragliu et al. 2011). ICTs are the means for knowledge and information exchange, opening the route to stakeholder collaboration (Jovicic 2019) and highlighting a relevant role and function for the Destination Management Organisation in innovation processes (Sorokina et al. 2022). Traditional arrangements of destination governance need to be transformed with the help of technologies facilitating exchange and collaboration towards adopting more sustainable approaches to destination development (Gretzel 2022).

A smart destination—whenever the ideal functioning is achieved—can generate efficiency for stakeholders in the destination and experience co-creation prior to innovative outcomes and sustainability (Gretzel et al. 2015; Trunfio and Campana 2019). As explained in the smart tourism conceptualisation, tourism innovation processes largely overlap with sustainability processes. Scholars need to work on these synergies, demonstrating that smartness and sustainability are not distinct agendas, yet highly interconnected, theoretically and in practice.

Second, the sustainability-smartness link has been addressed by *explicitly intertwining* the two dimensions with the *smart and sustainable destination*. This stream attempted to create explicit links between the two concepts, acknowledging the role of technologies in sustainable development. The sustainable city debate has been intertwined with the digital city, the information city, and smart city notions; all identifying technology as a driver of urban sustainable development (De Jong et al. 2015).

The smart city represents a decisive step in defining a smart and sustainable destination. The smart destination emerged from the evolution of the smart city concept (Camero and Alba 2019; Cavalheiro et al. 2019). Based on the different available definitions of a smart city (De Jong et al. 2015; Desdemoustier et al. 2019), attempts have been made to reframe the destination and the tourism imbalances within it by providing interpretative clues for dealing with overtourism issues (Pasquinelli and Trunfio 2020b).

Integration of the sustainable and smart destination concepts was discussed, especially in urban contexts where mobility, urban services, housing, and social segregation are deeply interconnected with the tourism system, recently fuelling interest in the overtourism phenomenon (Trunfio and Pasquinelli 2021). The debate is still taking its first steps, and there is a significant research gap concerning the smart destination approach to managing overcrowding and tourist overload (García-Hernández et al. 2019).

The Smart City Hospitality Framework (Koens et al. 2019) stated the centrality of a design-driven approach to governing tourism in the destination towards a sustainability transition. Local hospitality was framed by the following dimensions: liveability, experience quality, smart hospitality, sustainability, equitability (from natural, social, and economic perspectives), and resilience as the adaptation capacity of the local system to change. These are all integrated into the smart city that is pursuing sustainability (Koens et al. 2019).

A "smart tourism cities approach" to addressing overtourism was discussed in light of empirical findings concerning Spanish and Portuguese cities. The study stated the need to boost technology deployment in deeper integration with city management systems (Ivars-Baidal et al. 2019). Another empirical study highlighted the limited use of technological solutions to address overtourism, suggesting the priority goal of creating knowledge about the phenomenon amongst decision makers and the possibility of adopting different smart destination approaches based on destination specificities (García-Hernández et al. 2019).

The technological factor, in constant need for upgrades and improvements, was framed as the premise for smart and sustainable destination development. According to Shafiee et al. (2019), it is the fourth complementary pillar of sustainability, besides the environmental, economic, and social dimensions. There is a need to empirically investigate and test this four-pillar sustainable development model, to deal with the frictions characterising the "living" contexts where social and human dynamics and power relations may shed further tangible light on the path of sustainable development. Further empirical research may support policymakers in understanding how to build these four pillars of sustainability.

Smartness is a condition for achieving sustainability. This does not imply a prioritisation of technology over the drivers of sustainability. Instead, attention should be drawn to *technologies' contribution to knowledge-based and networking destinations*, holding together bottom-up processes stakeholder engagement processes with sustainable development and institutional top-down destination management (Pasquinelli and Trunfio 2020a). This view was complemented by a recent empirical study showing how destination stakeholders in two different Swedish destinations explained their understanding of a smart destination, not just in relation to technology deployment but through emphasising *values and beliefs guiding current and future change,* connected to their sense of the pursuit of sustainable development (Gelter et al. 2022).

3.4.2 *The Policy Domain: The EU Context and the European Capital of Smart Tourism*

Scholars have produced conceptual models interpreting the smart destination and its connections with sustainability. The policy domain at local, regional, national, and supranational (e.g. EU) levels assumes the combination of smartness and sustainability as a pillar for designing and financing policies and planning. The link between sustainability and smartness has been increasingly pivotal in practice and academic research. The European Union's Agenda on Growth and Jobs (i.e. the Europe 2020 strategy) promoted the entanglement of sustainability, smartness, and inclusiveness. For instance, this was translated into the tourism policy-oriented agenda through the third UNWTO Global Summit on City Tourism in December 2014, titled "New Paradigms in City Tourism Development". The statement of smartness potentially boosting inclusiveness and sustainability was a focal point, connecting with the United Nations New Urban Agenda and the 17 Sustainable Development Goals (UNWTO 2018).

The recent redefinition of the European Union's Urban Agenda confirmed the centrality of smartness and sustainability intersections concerning sustainable urban tourism. The Ex ante Assessment of the sustainable tourism thematic area under the Urban Agenda for the EU (Bellini 2022) identified the contribution of tourism to urban socio-economic development in relation to the development of smart cities and the digitalisation of urban services. Furthermore, it suggested the transversal role of digital technologies for "better regulation", "better knowledge", and "better funding".

Concerning "better regulation", the focus is on digital interoperability and data collection. Projects like the Smart City Control Rooms in Florence and Venice, and interregional networks such as the Tourism of Tomorrow Lab and Local Digital Twins promoting smart communities' data spaces demonstrate the relevance and efforts for developing digital data at the destination level. About "better knowledge", data "smart" use is highlighted with the efforts needed to transform data into the development of managerial tools and policies, turning them into a lever for collaborative governance. About "better funding", digitalisation for the green transition, services digitalisation, and SMEs digitalisation support are the main investment guidelines.

The European Capitals of Smart Tourism is an example of an initiative promoted by the European Commission aimed at "showcasing and supporting local practices in sustainable tourism" (Bellini 2022, p. 39). This framework encourages tailoring a local means of envisioning the entanglement of smartness and sustainability. Four award categories identified the four domains of smart tourism strategy-making, which are accessibility, digitalisation, sustainability, creativity, and cultural heritage.

According to a compendium of European Capitals of Smart Tourism best practices selected amongst the "innovative projects, initiatives and ideas" (2020, p. 8), sustainability "does not only mean to manage and protect your natural resources as a city, but to reduce seasonality and include the local community" (2020, p. 9). It is worth observing that sustainability can be understood as transversal to the other three domains (i.e. accessibility, digitalisation, and creativity and cultural heritage): (a) accessibility to travellers, regardless of age, social, and economic situation or disabilities, is an important component of inclusive tourism in line with the sustainability social pillar; (b) digitalisation, which includes digital mobility, digital accessibility, and digital skills improvement in local communities, is part of inclusive tourism and contributes to making tourism a lever for quality education, decent work, and economic growth; (c) creativity and cultural heritage concern the protection and enhancement of the local heritage to benefit contemporary cultural consumers (tourists and residents) and keep heritage alive and rejuvenated for future generations.

3.5 Technologies, Projects, and Practices for the Smart and Sustainable Destination

Although advances in academic investigation allowed framing new smart and sustainable scenarios, a focus on practice-oriented research is needed to exploit the technology opportunities for sustainable development. Applied research should connect the analysis of projects, practices, and tools with theoretical advancements informing practitioners' actions for smart and sustainable destinations. This research agenda might produce relevant managerial implications. First, it may demonstrate how "the smart and sustainable" can be framed together and take place in the destination and how many different configurations, contexts, and variables impact their intertwining. This may avoid the illusion that the mere adoption of technologies may contribute towards sustainability and can help local stakeholders to be more realistic and more imaginative in taking up the challenge to develop a smart and sustainable destination.

Second, there is a need to help policymakers and practitioners assess how and to what extent technologies contribute effectively to sustainability paths, considering all the criticalities and limitations of transitioning destinations. This need is particularly urgent given the significant technological investments supported by an extensive financing framework like the Next Generation EU in the European Union. Third, scholars need to contribute understanding and framing the bottlenecks that limit smart and sustainable destination development.

3.5.1 ICT Tools for the Smart and Sustainable Destination

ICT tools "require interpretation for [their] applications to be actionable" (Ali and Frew 2014b, p. 266). Their adoption can be "a practical approach" for destination managers to mitigate the negative impacts of tourism. However, there is a need to dig deeper into the multifaceted concept of sustainability and the local sustainable tourism vision to frame the potential of ICT tools in the broader framework of the smart and sustainable destination. In contrast with the broad technological focus in tourism literature, studies focusing on the exploitation of ICTs to support sustainable tourism development and management remained largely underdeveloped (Ali and Frew 2014b).

Ali and Frew's categorisation of technological tools supporting sustainable tourism development represents a starting point towards the understanding of ICTs' potential to contribute to sustainability (Ali and Frew 2014a). One of their merits consisted in drawing attention to the many smart sustainability tools, adding to the existing studies that focus on one single tool. This study will support the empirical effort in Chap. 4 to consider the different tools that sustainable tourism projects may rely on in the analysed destinations.

In Ali and Frew's categorisation, ICT-based tools were understood as those technologies used for "information processing which facilitate data processing, information sharing, communication, searching and selection" (Ali and Frew 2014a, p. 2).

The usage rationales of the ICT tools categorised by Ali and Frew (2014a) can be summarised as follows:

- *Data/information/knowledge production*: Collecting and combining data to support decision-making and information collection and sharing on the destination and tourism products for monitoring, evaluating, and forecasting (trends and resource consumption).
- *Fostering stakeholder engagement and participation*: Sharing and connecting for stakeholders' participation, engagement, and partnership building.
- *Destination experience delivery/co-creation*: Communicating, marketing, and shaping the destination experience, with an orientation to the diverse internal and external stakeholders.

Based on Ali and Frew's definitions of the different ICT tools (see Chap. 2 for the definitions), Fig. 3.1 groups the tools based on their primary usage rationale identified by the authors. However, the overlapping areas in the diagram suggest the possible co-existence of diverse usage rationales. The authors did not highlight these overlapping areas. They depend upon the exploitation of their full potential and their integration into the broader strategic framework and sustainability vision guiding their deployment (see Chap. 4).

Ali and Frew (2014b) explored the usage of ICT tools by investigating destination managers' first-hand professional experience, providing insights into ICT adoption. The findings revealed that over one-third of respondents did not use ICTs for sustainable development actions or used them to a limited extent. Destination management systems were the most indicated, followed by intelligent transport systems and environmental management information systems (see Chap. 2 for the definitions). Computer simulation and virtual tourism were the lowest mentioned tools in the produced ranking.

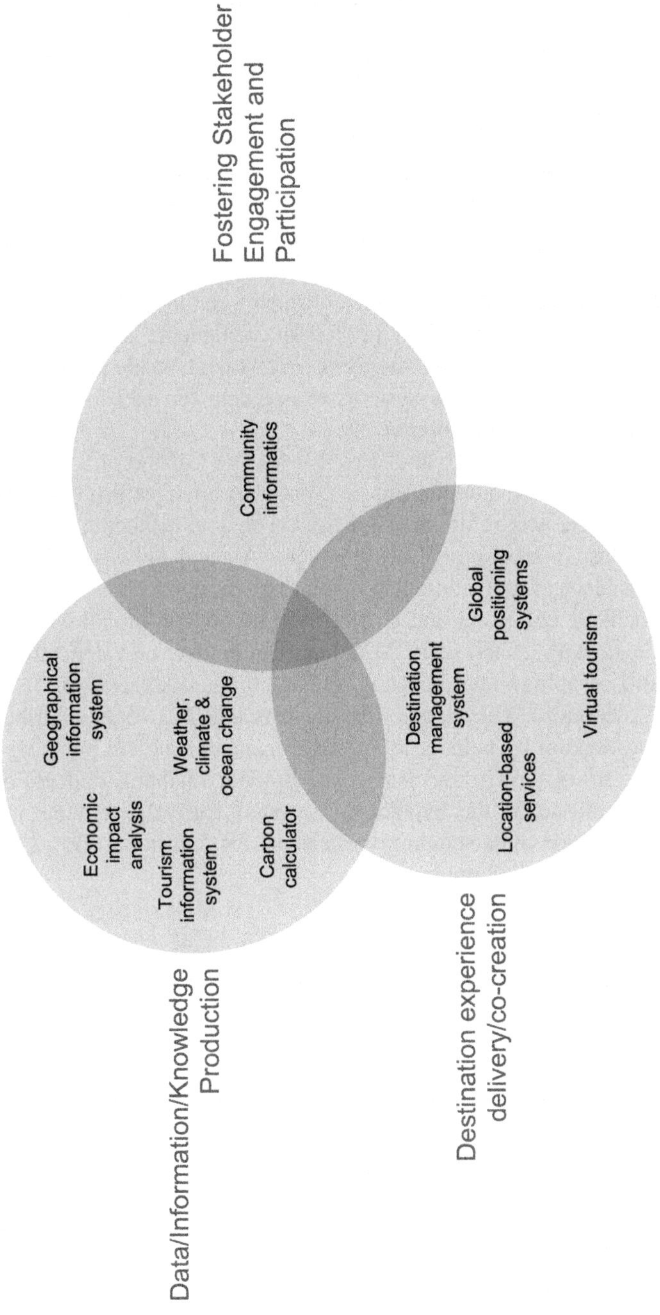

Fig. 3.1 ICT tools based on Ali and Frew's categorisation (Ali and Frew 2014a)

Based on a different methodology but relying on Ali and Frew's categorisation, Calza et al. (2022) investigated the ICT tools adopted for sustainable tourism development in the international and Italian contexts. In the sample of international destinations, virtual tourism (31%), location-based services (19%), and community informatics (17%) emerged as the key tools and platforms adopted to address and manage tourism imbalances and impacts. Similar technical attributes characterise virtual tourism and location-based services, such as location-based games, augmented reality, and sensors, shaping the on-site destination experience. Community informatics is based on exploiting Internet platforms and virtual spaces to assist residents' and tourists' engagement with and exploration of the destination (e.g. WeChat Helsinki and MyHelsinki). In the analysed Italian context, virtual tourism and community informatics are confirmed to be the most relevant, together with a tourism information system (27%) for sustainable tourism development (Calza et al. 2022). This witnesses the recent investments made in data warehouses for the evident need for tourism systems to use data for decision-making, at the individual business and destination levels.

Although Ali and Frew's (Ali and Frew 2014b) and Calza et al.'s (2022) findings are not comparable for several reasons, a hypothesis emerges from their juxtaposition. Over time, there was a shift, or at least an expansion, in the modality to envision ICT tools for sustainable tourism development. Ali and Frew's analysis in 2014 found the prevalence of tools for managing tourism products, transport systems, and the environment by collecting and sharing information (information-oriented ICT adoption). Calza et al.'s analysis in 2022 found an evident prevalence of ICT tools for enhancing, shaping, and boosting the destination experience (experience-oriented ICT adoption). This hypothesis suggests that ICT tools contribute to the sustainable destination by helping the rational management and planning of products, services, flows, and movements and digitally enabling and enriching the destination experience. If the hypothesis is tested correctly, a relevant research agenda would include understanding the reasons for expanding the ICT adoption approaches, the effects, and implications.

3.5.2 *Interpreting ICT Tool Applications in the Frame of Sustainable Tourism Policies and Actions*

Beyond the ICT tools description and the study of characteristics and adoption, the efforts to interpret and place them within the broader frame of policies and actions for sustainable tourism remain limited. In the sustainable tourism debate, recent contributions to overtourism drew attention to using technologies (García-Hernández et al. 2019; Ivars-Baidal et al. 2019; Trunfio and Pasquinelli 2021). The adoption of ICT tools emerged as being limited in scope (Ivars-Baidal et al. 2019), without clarity on its positioning in the overall strategy to address tourism imbalances. This might be the reason for the limited capacity of these tools to effectively push the destination towards a more sustainable development trajectory (García-Hernández et al. 2019).

Theoretical efforts to conceptualise tourism through the smart city highlighted the role of ICT tools in boosting sustainability. The Smart City Hospitality framework (Koens et al. 2019) and the Smart City Lens (Pasquinelli and Trunfio 2020a, b) made this point, arguing that ICT tools may be a decisive support for residents', entrepreneurs', workers', and tourists' engagement. Empirical analysis, however, is needed to demonstrate how the role of ICT tools that was highlighted in theoretical models can be played in practice. Trunfio and Pasquinelli (2021) highlighted smart tech-based and traditional actions for shaping tourists' flows and behaviours towards a more sustainable destination development. Visitor management has been increasingly considered a fundamental lever of sustainable destination management (Kebete and Wondirad 2019).

Smart-tech-based actions are built around tool applications, which emerged as the enabling factor. Some examples are travelling mobile tracking systems, big data analysis, forecasting, and tracking systems to analyse behaviours and targeting, smart ticketing, dynamic pricing/discounts, and mobile gaming apps. Four main strategic uses of ICT tools emerged, such as tourist flows dispersion, building destination marketing plans, integration of tourism and urban planning, and visitor engagement.

The Smart Technology Matrix (Trunfio and Pasquinelli 2021) helped further explain the use of ICT tools. This was composed of two dichotomies: the mono-dimensional logic of adoption versus the integrated logic of adoption; and top-down versus bottom-up adoption. The first dichotomy concerns the *extension* of the sustainability-oriented ICT tool usage. It can be extended only to tourists and tourism stakeholders (mono-dimensional logic of intervention) or encroach on the tourism industry borders, including the local community, workers, and entrepreneurs who are indirectly connected with the tourism sectors (integrated logic of intervention). The second dichotomy concerns the *direction* of the sustainability-oriented action. If supporting centralised destination decision-making, planning and action implementation, the ICT tool is categorised as top-down intervention; if supporting dispersed decision-making, planning, grassroots actions, and responsible behaviours, the ICT tool is categorised as bottom-up intervention.

For instance, ICT tools to define and set access limits at the destination are positioned at the intersection of the top-down and mono-dimensional logic of intervention, as a regulative action, whose enforcement can be supported by technology. ICT platforms for integrated city management, monitoring and collecting data on mobility, housing, public space access, and crowding were categorised at the intersection of the integrated logic of intervention and top-down intervention. ICT tools deployment for boosting commitment and knowledge sharing amongst tourism intermediaries (e.g. tour operators and tourist guides) were at the intersection between mono-dimensional logic and bottom-up intervention. Smart ticketing, dynamic pricing systems, and mobile gaming apps for engaging city users filled the intersection between the integrated logic of intervention and bottom-up intervention.

Figure 3.2 presents a framework helping to design the empirical analysis and interpretating the findings, starting from the observation of the ICT tools, their description, and understanding of their adoption rationale.

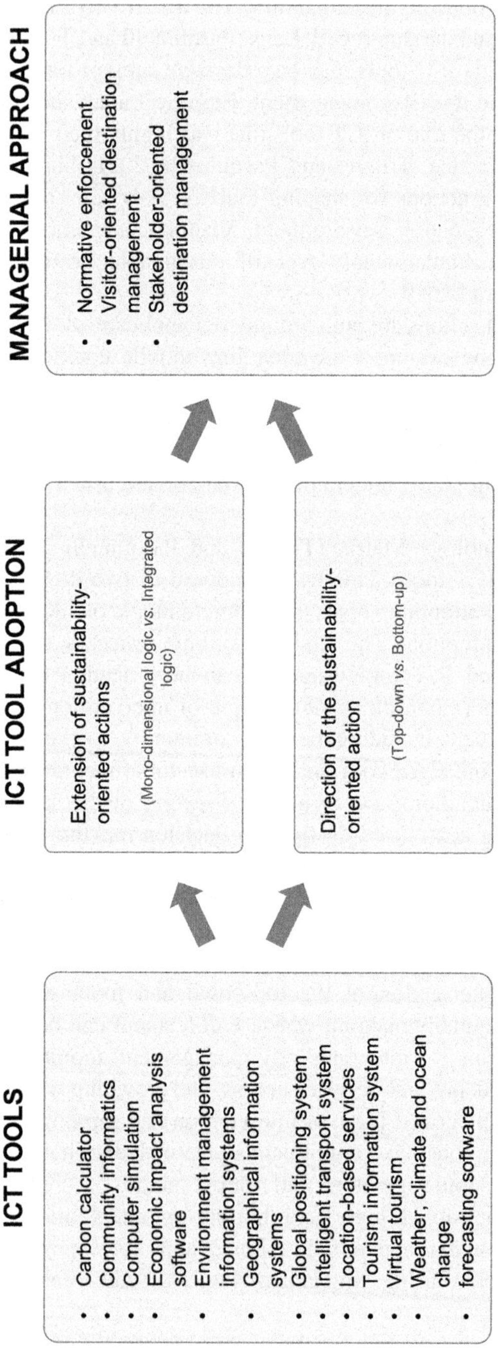

Fig. 3.2 ICT tools, ICT adoption, managerial approaches: An analytical framework. Source: The authors; developed from Ali and Frew 2014a

3.6 Towards an Analytical Framework for Smart and Sustainable Destination Management

This section builds on the previous literature overview and discussion and presents an analytical framework connecting the ICT tools' adoption with the intervention approaches, strategies, and sustainable tourism visions. Figure 3.3 represents this analytical framework where ICTs are the outer layer (the top of the pyramid), which is the starting point for exploring the smart and sustainable destination.

This analytical framework helps us discuss actions in relation to a substratum of strategic frames and sustainable tourism vision. It deepens insight into the rationale and implications of actions and assists the design of action evaluation frameworks. Connecting the single action with a broader strategic framework and sustainable tourism vision suggests how to monitor and measure the impacts of the actions undertaken and supports the destination community towards the definition of tourism development goals.

ICT tools, framed within actions, are the top layer of a pyramid visible through destination stakeholders' decision-making and behaviours. Beneath this, a second pyramid layer concerns the *strategic framework* in which the actions may take place. The sustainable destination strategy can be a deliberate or an emerging development strategy. In both cases, three strategic frameworks are identified: the *conservative*, the *radical*, and the *transition frameworks* for tourism systems. The conservative and radical frameworks were discussed in this chapter as emerging from the existing literature. In contrast, the transition framework is meant to complete the array of

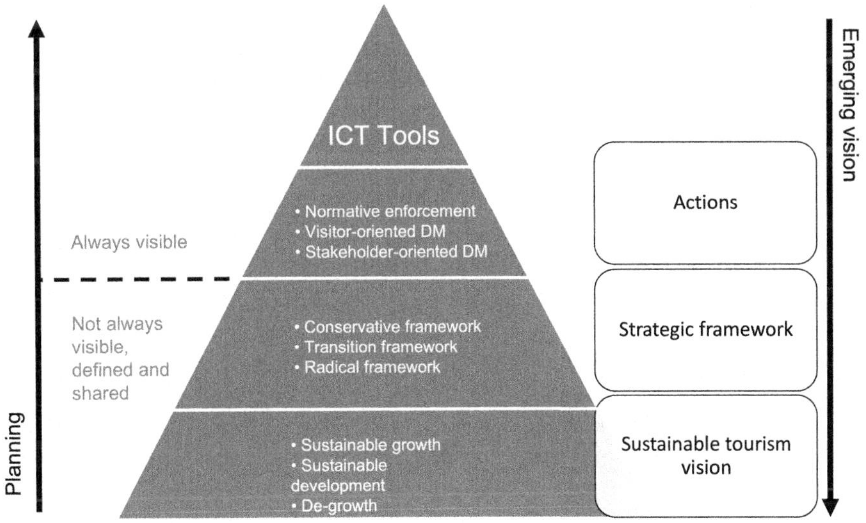

Fig. 3.3 Analytical framework: ICT tools, actions, strategic frameworks, and sustainable tourism visions

possible strategic scenarios by introducing a middle-ground strategy between the conservative and the radical ones.

This highlights the incremental steps that make sustainability enter the destination development agenda in a smooth way. Based on Lindblom's notion of incrementalism (Lindblom 1959), single actions are derived from the agreements reached amongst the relevant actors, also considering contingent opportunities, which solidify over time into a sustainable destination strategy without this being preordained to the single action. A slow path introducing only marginal progress towards sustainability characterises this strategic framework. It is defined as the "middle ground" approach (Hudson 1992) because, although it responds to a sustainable tourism vision, it is meant to be "workable" as it does not connect the single action (and its assessment) directly to the "big" goals that this vision implies. The three strategic frameworks presented are meant to cover all the possible scenarios, but all present limitations and pitfalls that future empirical and conceptual research should reveal in detail and help to discuss.

Beneath this layer is the *sustainable tourism vision* adopted as the "big" goal of strategic frameworks and actions. The visions are *sustainable growth, sustainable development*, and *the de-growth imperative*. The sustainable growth vision combines sustainability with economic growth, which is conceived as the irrevocable foundation of economically sustainable tourism. It maintains "growth as usual" from a quantitative perspective as compatible with pursuing sustainability goals. Sustainable development refers to the expansion of the qualitative content of tourism products and destinations and prioritises this improvement over the quantitative growth of economic flows. For instance, it seeks sustainability through an expansion of tourists' expenditure in contrast with boosting the growth of incoming tourism flows. It denies a priori that growth and sustainability can be compatible and "emphasise[s] the need to consider qualitative content of growth to reduce [tourism] material and energy intensity and inequitable impacts as well as to promote reforms to maintain stocks of ecological capital, address inequalities in the distribution of income and reduce vulnerability to crises" (Pike et al. 2017, p. 138). Third, the "degrowth imperative" (Higgins-Desbiolles et al. 2019; Cheung and Li 2019) postulates that a decrease in the quantitative and economic tourism flows is unavoidable if sustainability is to be pursued effectively. This resonates with "transformative tourism" for radically reshaping the tourism development model (Milano et al. 2019).

Sustainable destination *planning* emerges if ICT tools and actions emanate from a defined strategic framework to pursue a defined sustainability vision. However, ICT tools and actions may be put in place without a clear strategic framework and outside of any sharp sustainable tourism vision. In this case, actions are likely to be followed by a series of—even contrasting—actions, responding only to contingent needs and individual stakeholder mindsets. In other cases, actions—even though designed outside of any strategic framework—can shape an *emerging sustainable tourism vision* for the destination community in the long run.

A mix of actions belonging to the different described categories can be implemented. Deliberately or not (i.e. based on the presence or absence of a clearly defined sustainable tourism vision), normative enforcement, visitor- and stakeholder-oriented destination management can have different implications. Let us take the example of limited access action (i.e. normative enforcement). It may be framed as a conservative strategy if the limit serves to deny access only in case of extreme and intolerable incoming visitor flows, defined as being on the level of tourists' toleration of crowding, rather than on the local community's needs. It may be a short-term action adopted in specific local contingencies (short-term) or simply limiting physical access to a specific tourism precinct, without considering the general implications for local tourism sustainability (e.g. ending up with a growing load on other destination sites where negative effects can be even more disruptive). This action can be part of a conservative strategic framework aligned with a sustainable growth vision. In contrast, depending on the definition and implementation modality of access limitations (e.g. structurally reducing cruise tourist inflows in the frame of a broader regional policy), this action can serve as a transformative lever in the frame of a radical strategy towards tourism de-growth.

References

Addis M, Holbrook MB (2006) On the conceptual link between mass customisation and experiential consumption: an explosion of subjectivity. J Consum Behav 1(1):50–66

Ali A, Frew AJ (2014a) ICT and sustainable tourism development: an innovative perspective. J Hosp Tour Technol 5(1):2–16

Ali A, Frew AJ (2014b) Technology innovation and applications in sustainable destination development. Inf Tech Tour 14(4):265–290

Andereck KL, Valentine KM, Knopf RC, Vogt CA (2005) Residents' perceptions of community tourism impacts. Ann Tour Res 32(4):1056

Andersson G (2017) Visitor streams in city destinations: towards new tools for measuring urban tourism. In: Bellini N, Pasquinelli C (eds) Tourism in the city. Towards an integrative agenda on urban tourism. Springer

Andreoli A, Cuccu O, Silvestri F (2017) Il turismo come opportunità di sviluppo per le aree interne del Paese. In: Becheri E, Micera R, Morvillo A (eds) XXI Rapporto sul Turismo Italiano. Rogiosi Editore, Napoli

Ap J (1992) Residents' perceptions on tourism impacts. Ann Tour Res 19:665–690

Arbolino R, Boffardi R, De Simone L, Ioppolo G (2021) Multi-objective optimization technique: a novel approach in tourism sustainability planning. J Environ Manag 285(May):112016

Asmelash AG, Kumar S (2019) Assessing progress of tourism sustainability: developing and validating sustainability indicators. Tour Manag 71(April):67–83

Bathelt H, Malmberg A, Maskell P (2004) Clusters and knowledge: local buzz, global pipelines and the process of knowledge creation. Prog Hum Geogr 28(1):31–56. https://doi.org/10.1191/0309132504ph469oa

Bailey C, Miles S, Stark P (2004) Cultureled urban regeneration and the revitalisation of identities in Newcastle, Gatehead and the North East of England. Int J Cult Policy 10(1):46–65

Bellini N (2022) Ex-ante assessment of the sustainable tourism thematic area under the urban agenda for the EU. Final Report. European Commission.

Bellini N, Go FM, Pasquinelli C (2017) Urban tourism and city development: notes for an integrated policy agenda. In: Tourism in the City. Springer, Cham, pp 333–339

Beritelli P (2011) Cooperation among prominent actors in a tourist destination. Ann Tour Res 38(2):607–629

Beritelli P (2019) Transferring concepts and tools from other fields to the tourist destination: a critical viewpoint focusing on the lifecycle concept. J Destin Mark Manag 14:100384

Blancas FJ, Lozano-Oyola M, Gonzàlez M, Caballero R (2016) Sustainable tourism composite indicators: a dynamic evaluation to manage changes in sustainability. J Sustain Tour 24(10): 1403–1424

Bowden JL-H (2009) The process of customer engagement: a conceptual framework. J Mark Theory Pract 17(1):63–74. https://doi.org/10.2753/MTP1069-6679170105

Bramwell B, Lane B (2011) Critical research on the governance of tourism and sustainability. J Sustain Tour 19(4–5):411–421

Brida JG, Cortes-Jimenez I, Pulina M (2014) Has the tourism-led growth hypothesis been validated? A literature review. Curr Issue Tour 19(5):1–37

Brodie RJ, Ilic A, Juric B, Hollebeek L (2013) Consumer engagement in a virtual brand community: an exploratory analysis. J Bus Res 66(1):105–114. https://doi.org/10.1016/j.jbusres.2011.07.029

Burrai E, Buda DM, Stanford D (2019) Rethinking the ideology of responsible tourism. J Sustain Tour 27(7):992–1007

Butler RW (1999) Sustainable tourism: a state-of-the-art review. Tour Geogr 1(1):7–25

Calza F, Tunfio M, Pasquinelli C, Sorrentino A, Campana S, Rossi S (2022) Technology-driven innovation. Exploiting ICTs tools for digital engagement, smart experiences and sustainability in tourism destinations. SLIOB. Enzo Albano Edizioni, Naples

Camero A, Alba E (2019) Smart City and information technology: a review. Cities 93, October:84–94

Campòn-Cerro AM, Hernàndez-Mogollòn JM, Alves H (2017) Sustainable improvement of competitiveness in rural tourism destinations: the quest for tourist loyalty in Spain. J Destin Mark Manag 6(3):252–266

Caragliu A, Del Bo C, Nejkamp P (2011) Smart cities in Europe. J Urban Technol 18(2):65–82

Carayannis EG, Campbell DF (2014) Developed democracies versus emerging autocracies: arts, democracy, and innovation in quadruple helix innovation systems. J Innov Entrepreneurship 3(1):201–234

Cavalheiro MB, Joia LA, Cavalheiro GM (2019) Towards a smart tourism destination development model: promoting environmental, economic, socio-cultural and political values. Tour Plan Develop 17(3):1–23

Cheer JM, Milano C, Novelli M (2019) Tourism and community resilience in the Anthropocene: accentuating temporal overtourism. J Sustain Tour 27(4):554–572

Cheer JM, Lapointe D, Mostafanezhad M, Jamal T (2021) Global tourism in crisis: frameworks for research and practice. J Tour Future 7(3):278–294

Cheung KS, Li LH (2019) Understanding visitor–resident relations in overtourism: developing resilience for sustainable tourism. J Sustain Tour 27(8):1197–1216

Colazzo L, Molinari A, Villa N (2008) From e-learning to ":Co-learning": the role of virtual communities BT - learning to live in the knowledge society (M. Kendall & B. Samways (eds.); pp. 329–338). Springer.

Colomb C, Novy J (2016) Protest and resistance in the tourist city. Routledge

Comunian R, Sacco PL (2006) NewcastleGateshead: riqualificazione urbana e limiti della citta creativa. WP Universita Iuav di Venezia, Venice

CREST (2018) The case for responsible travel: trends & statistics 2018. Centre for Responsible Travel

Crouch GI (2011) Destination competitiveness: an analysis of determinants attributes. J Travel Res 50(1):27–45

De Jong M, Joss S, Schraven D, Zhan C, Weijnen M (2015) Sustainable-smart-resilient-low carbon-eco-knowledge cities; making sense of a multitude of concepts promoting sustainable urbanization. J Clean Prod 109:25–38

De Neve J-E, Sachs JD (2020) The SDGs and human Well-being: a global analysis of synergies, trade-offs, and regional differences. Sci Rep Nat Res 2020(10–15113):1–12

Desdemoustier J, Crutzen N, Cools M, Teller J (2019) Smart City appropriation by local actors: an instrument in the making. Cities 92(September):175–186
Diaz-Parra I, Jover J (2021) Overtourism, place alienation and the right to the city: insights from the historic Centre of Seville, Spain. J Sustain Tour 29(2–3):158–175
Dodds R, Butler RW (2019) Introduction. In: Dodds R, Butler RW (eds) Overtourism. Issues, realities and solutions. De Gruyter, Berlin
Donaldson T, Preston LE (1995) The stakeholder theory of the corporation: concepts, evidence, and implications. Acad Manag Rev 20(1):65–91
Dredge D (2017) "Overtourism" Old wine in new bottles? Retrieved December 8, 1BC. https://www.linkedin.com/pulse/overtourism-old-wine-new-bottles-dianne-dredge. Accessed 11 Jan 2020
Dumbrovská V (2017) Urban tourism development in Prague: from tourist mecca to tourist ghetto. In: Tourism in the city. Springer, Cham, pp 275–283
Dunn A (2005) Structures of radio drama. In: Fulton H, Huisman R, Murphet J, Dunn A (eds) Narrative and media. Cambridge University Press, New York, pp 191–202
Dwyer L (2018) Saluting while the ship sinks: the necessity for tourism paradigm change. J Sustain Tour 26(1):29–48
Dwyer L, Kim C (2003) Destination competitiveness: determinants and indicators. Curr Issue Tour 6(5):364–414
Eshuis J, Klijn E-H, Braun E (2014) Place marketing and citizen participation: branding as strategy to address the emotional dimension of policy making? Int Rev Adm Sci 80(1):151
Evans G (2003) Hard-branding the cultural city–from Prado to Prada. Int J Urban Reg Res 27(2): 417–440
Farsari I (2021) Exploring the nexus between sustainable tourism governance, resilience and complexity research. Tour Recreat Res:*1*
Fava N, Rubio SP (2017) From Barcelona: the pearl of the Mediterranean to bye bye Barcelona. In: Bellini N, Pasquinelli C (eds) Tourism in the City. Towards an integrative agenda on urban tourism. Springer, pp 285–296
Festa G, Riad Shams SM, Metallo G, Cuomo T (2020) Enhancing stakeholder networks in wine tourism–Evidence from Italian small municipalities. EuroMed J Bus 15(3):349–360
Fleishman L, Feitelson E, Salomon I (2004) The role of cultural and demographic diversity in crowding perception: evidence from nature reserves in Israel. Tour Anal 9(1–2):23–40
Font X, McCabe S (2017) Sustainability and marketing in tourism: its contexts, paradoxes, approaches, challenges and potential. J Sustain Tour 25(7):869–883
Franch M, Martini U, Buffa F, Parisi G (2008) 4L (landscape, leisure, learning and limit): responding to new motivations and expectations of tourists to improve the competitiveness of alpine destinations in a sustainable way. Tour Rev 63(1):4–14
Franklin A (2016) Journeys to the Guggenheim museum Bilbao: towards a revised Bilbao effect. Ann Tour Res 59(July):79–92
Freeman RE (1984) Strategic management: a stakeholder approach. Pitman, Boston
Freudenreich B, Lüdeke-Freund F, Schaltegger S (2020) A stakeholder theory perspective on business models: value creation for sustainability. J Bus Ethics 166(1):3–18
Frey N, George R (2010) Responsible tourism management: the missing link between business owners' attitudes and behaviour in the Cape Town tourism industry. Tour Manag 31:621–628
Füller H, Michel B (2014) "Stop being a tourist!" new dynamics of urban tourism in Berlin-Kreuzberg. Int J Urban Reg Res 38(4):1304–1318
Fulton H (2005) Print news as narrative. In: Fulton H, Huisman R, Murphet J, Dunn A (eds) Media and narrative. Cambridge University Press, New York, pp 218–244
García-Hernández M, Ivars-Baidal J, Mendoza de Miguel S (2019) Overtourism in urban destinations: the myth of smart solutions. Boletin de la Asociacion de Geografos Espanoles 83:1–38
Garrod B, Fyall A (1998) Beyond the rhetoric of sustainable tourism? Tour Manag 3(19):199–212
Gartner WC (1994) Image formation process. J Travel Tour Mark 2(2–3):191–216

Gelter J, Fuchs M, Lexhagen M (2022) Making sense of smart tourism destinations: a qualitative text analysis from Sweden. J Destin Mark Manag 23:100690
Go FM, Trunfio M (2011) E-services governance in public and private sectors: a destination management organization perspective. In: D'Atri A, Ferrara M, George JF, Spagnoletti P (eds) Information technology and innovation trends in organizations. Springer, pp 11–19
Goodwin H (2019) Barcelona–crowding out the locals: a model for tourism management? In: Dodds R, Butler RW (eds) Overtourism. Issues, realities and solutions. De Gruyter, Berlin, pp 125–138
Gotham KF (2002) Marketing Mardi Gras: commodification, spectacle and the political economy of tourism in New Orleans. Urban Stud 39(10):1735–1756
Govers R, Go F (2009) Place branding. Glocal, virtual and physical identities, constructed, imagined and experienced. Palgrave
Gravari-Barbas M, Guinand S (2017) Tourism and gentrification in contemporary metropolises: international perspectives. Routledge, London, UK
Greenwood M (2007) Stakeholder engagement: beyond the myth of corporate responsibility. J Bus Ethics 74:315–327
Gretzel U (2022) The smart DMO: a new step in the digital transformation of destination management organisations. Euro J Tour Res 30:302
Gretzel U, Sigala M, Xiang Z, Koo C (2015) Smart tourism: foundations and developments. Electron Mark 25:179–188
Gutiérrez-Taño D, Garau-Vadell JB, Díaz-Armas RJ (2019) The influence of knowledge on residents' perceptions of the impacts of overtourism in P2P accommodation rental. Sustainability (Switzerland) 11(4)
Gui L, Russo AP (2011) Cruise ports: a strategic nexus between regions and global lines—evidence from the Mediterranean. Marit Policy Manag 38(2):129–150
Hall CM (2008) Tourism planning: policies, processes and relationships. Pearson
Hall MC (2003) Tourism issues, agenda setting and the media. e-Rev Tour Res 1(3):42–45
Hall MC (2011) Policy learning and policy failure in sustainable tourism governance: from first- and second-order to third-order change? J Sustain Tour 19(4–5):649–671
Hall CM, Mitchell R (2000) Wine tourism in the Mediterranean: a tool for restructuring and development. Thunderbird Int Bus Rev 42(4):445–465
Hedrick-Wong Y, Choong D (2014) Mastercard 2014 global destination cities index. Master Card Worldwide Insights, Mastercard
Hernandez S (2006) Striving for control: democracy and oligarchy at a Mexican cooperative. Econ Ind Democr 27(1):105–135
Higgins-Desbiolles F, Carnicelli S, Krolikowski C, Wijesinghe G, Boluk K (2019) Degrowing tourism: rethinking tourism. J Sustain Tour 27:1926. https://doi.org/10.1080/09669582.2019.1601732
Hörisch J, Freeman RE, Schaltegger S (2014) Applying stakeholder theory in sustainability management: links, similarities, dissimilarities, and a conceptual framework. Organ Environ 27(4):328–346. https://doi.org/10.1177/1086026614535786
Hudson B (1992) Community care planning: incrementalism to rationalism? Soc Policy Adm 26(3): 185–200
Hueting R, Reijnders L (1998) Sustainability is an objective concept. Ecol Econ 27:139–147
Ivars-Baidal JA, Celdrán-Bernabeu MA, Mazón JN, Perles-Ivars ÁF (2019) Smart destinations and the evolution of ICTs: a new scenario for destination management? Curr Issue Tour 22(13): 1581–1600
Jakob D (2013) The eventification of place: urban development and experience consumption in Berlin and new York City. European Urban and Regional Studies 20(4):447–459
Jin Q, Hu H, Kavan P (2016) Factors influencing perceived crowding of tourists and sustainable tourism destination management. Sustainability (Switzerland) 8:1–17
Joppe M (2018) Tourism policy and governance: quo vadis? Tour Manag Perspect 25:201–204

Jovicic DZ (2019) From the traditional understanding of tourism destination to the smart tourism destination. Curr Issue Tour 22(3):276–282
Kavaratzis M, Hatch MJ (2021) The elusive destination brand and the ATLAS wheel of place Brand Management. J Travel Res 60(1):3–15
Klijn EH, Eshuis J, Braun E (2012) The influence of stakeholder involvement on the effectiveness of place branding. Public Manag Rev 14(4):499–519
Kebete Y, Wondirad A (2019) Visitor management and sustainable destination management nexus in Zeige peninsula, northern Ethiopia. J Destin Mark Manag 13(September):83–98
Kim D, Lee CK, Sirgy MJ (2016) Examining the differential impact of human crowding versus spatial crowding on visitor satisfaction at a festival. J Travel Tour Mark 33(3):293–312
Kim S, Kim N (2020) A social cost-benefit analysis of the vehicle restriction policy for reducing overtourism in Udo, Korea. Sustain 12(2):612
Koens K, Melissen F, Mayer I, Aall C (2019) The Smart City hospitality framework: creating a foundation for collaborative reflections on overtourism that support destination design. J Destin Mark Manag 19:100376. https://doi.org/10.1016/j.jdmm.2019.100376
Koens K, Postma A, Papp B (2018) Is overtourism overused? Understanding the impact of tourism in a city context. Sustainability (Switzerland) 10:1–15
Laesser C, Beritelli P (2013) St. Gallen consensus on destination management. J Destin Mark Manag 2(1):46–49. https://doi.org/10.1016/j.jdmm.2012.11.003
Laplume AO, Sonpar K, Litz RA (2008) Stakeholder theory: reviewing a theory that moves us. J Manag 34(6):1152–1189
Laureti LL, Rogges M, G., Costantiello, A. (2018) Evaluation of economic, social effects of renewable energy technologies. J Environ Prot 9:1143–1154
Lavandoski J, Vargas-Sànchez A, Silva JA (2018) Causes and effects of wine tourism development in organisational context: the case of Alentejo, Portugal. Tour Hosp Res 18(1):107–122
Leydesdorff L, Etzkowitz H (1996) Emergence of a triple helix of university—Industry—Government relations. Sci Public Policy 23(5):279–286
Lindberg K, McCool S, Stankey G (1997) Rethinking carrying capacity. Ann Tour Res 24(2): 461–465. https://doi.org/10.1016/S0160-7383(97)80018-7
Lindblom CE (1959) The science of "muddling through". Public Adm Rev 19(2):79–88
Liu Z (2003) Sustainable tourism development: a critique. J Sustain Tour 11(6):459–475
Loureiro SM, Romero J, Bilro RG (2020) Stakeholder engagement in co-creation process for innovation: a systemic literature review and case study. J Bus Res 119(October):388–409
Lüdeke-Freund F, Dembek K (2017) Sustainable business model research and practice: emerging field or passing fancy? J Clean Prod 168:1668–1678. https://doi.org/10.1016/j.jclepro.2017.08.093
Martín JMM, Martínez JMG, Fernández JAS (2018) An analysis of the factors behind the citizen's attitude of rejection towards tourism in a context of overtourism and economic dependence on this activity. Sustainability (Switzerland) 10(8). https://doi.org/10.3390/su10082851
McCool SF (1994) Planning for sustainable nature dependent tourism development: the limits of acceptable change system. Tour Recreat Res 19(2):51–55
McCool SF, Lime DW (2001) Tourism carrying capacity: tempting fantasy or useful reality? J Sustain Tour 9(5):372–388
Milano C (2017) Overtourism y Turismofobia. Tendencias globales y contextos locales. Ostelea School of Tourism & Hospitality, Barcelona
Milano C, Novelli M, Cheer JM (2019) Overtourism and Tourismphobia: a journey through four decades of tourism development, planning and local concerns. Tour Plan Develop 16(4): 353–357
Minoia P (2017) Venice reshaped? Tourist gentrification and sense of place. In: Bellini N, Pasquinelli C (eds) Tourism in the city. Towards an integrative agenda on urban tourism. Springer, pp 261–274

Muler Gonzalez V, Coromina L, Galí N (2018) Overtourism: residents' perceptions of tourism impact as an indicator of resident social carrying capacity–case study of a Spanish heritage town. Tour Rev 73(3):277–296. https://doi.org/10.1108/TR-08-2017-0138

Neuts B, Nijkamp P (2012) Tourist crowding perception and acceptability in cities. An applied modelling study on Bruges. Ann Tour Res 39(4):2133–2153

Novy J, Colomb C (2019) Urban tourism as a source of contention and social mobilisations: a critical review. Tour Plan Develop 16:358. https://doi.org/10.1080/21568316.2019.1577293

O'Brian D, Miles S (2010) Cultural policy as rhetoric and reality: a comparative analysis of policy making in the peripheral north of England. Cultural Trends 19(1/2):3–13

OECD (2016) OECD tourism trends and policies 2016. OECD Publishing

Oklevik O, Gössling S, Michael CH, Jens Kristian SJ, Ivar Petter G, McCabe S (2019) Overtourism, optimisation, and destination performance indicators: a case study of activities in fjord Norway. J Sustain Tour 27(12):1804–1824

Ooi CD (2004) Poetics and politics of destination branding: Denmark. Scand J Hosp Tour 4(2): 107–121

Pásková M, Wall G, Zejda D, Zelenka J (2021) Tourism carrying capacity reconceptualization: modelling and management of destinations. J Destin Mark Manag 21:100638

Pasquinelli C (2014) Branding as urban collective strategy-making: the formation of Newcastle Gateshead's organizational identity. Urban Stud 51(4):727–743

Pasquinelli C, Bellini N (2017) Global context, policies and practices in urban tourism: an introduction. In: Bellini N, Pasquinelli C (eds) Tourism in the city. Towards an integrative agenda on urban tourism. Springer, pp 1–28

Pasquinelli C, Maiello A (2015) Destruction or construction? A (counter) branding analysis of sport mega-events in Rio de Janeiro. Cities 48(November):116–124

Pasquinelli C, Trunfio M (2020a) Overtouristified cities: an online news media narrative analysis. J Sustain Tour 28(11):1805–1824

Pasquinelli C, Trunfio M (2020b) Reframing urban Overtourism through the smart-city lens. Cities 102:1–8

Pasquinelli C, Trunfio M (2020c) The European capitals of culture in a time of overtourism hazard. In: Cantoni L, De Ascaniis S, Elgin-Nijhuis K (eds) Proceedings of the Heritage, Tourism and Hospitality International Conference HTHIC2020. Università della Svizzera Italiana, Lugano, pp 159–170

Pasquinelli C, Trunfio M (2022) The missing link between overtourism and post-pandemic tourism. Framing twitter debate on the Italian tourism cities. J Place Manag Dev 15(3):229–247

Pasquinelli C, Trunfio M, Bellini N, Rossi S (2022) Reimagining urban destination: adaptive and transformative city brand attributes and values in the pandemic crisis. Cities 124(May):103621

Passetti E, Bianchi L, Battaglia M, Frey M (2019) When democratic principles are not enough: tensions and temporalities of dialogic stakeholder engagement. J Bus Ethics 155(1):173–190. https://doi.org/10.1007/s10551-017-3500-z

Pavlić I, Portolan A (2016) In: Jafari J, Xiao H (eds) Irritation index BT–encyclopedia of tourism. Springer International Publishing, p 495. https://doi.org/10.1007/978-3-319-01384-8_564

Peeters, P. M., Gössling, S., Klijs, J., Milano, C., Novelli, M., Dijkmans, C. H. S., ... Postma, A. (2018). Research for TRAN committee-Overtourism: impact and possible policy responses. European Parliament, Policy Department for Structural and Cohesion Policies.

Perdue R, Long PT, Kang YS (1999) Boomtown tourism and resident quality of life: the marketing of gaming to host community residents. J Bus Res 44(3):165–177

Phillips R (1997) Stakeholder theory and a principle of fairness. Bus Ethics Q 7(1):51–66

Pike A, Rodriguez-Pose A, Tomaney J (2017) Local and regional development, 2nd edn. London, Routledge

Pizam A (1978) Tourism's impacts: the social costs to the destination community as perceived by its residents. J Travel Res 16(4):8–12

Postma A, Schmuecker D (2017) Understanding and overcoming negative impacts of tourism in city destinations: conceptual model and strategic framework. J Tour Future 3(2):144–156

Punzo G, Trunfio M, Castelllano R, Buonocore M (2022) A multi-modelling approach for assessing sustainable tourism. Soc Indic Res 163:1399. https://doi.org/10.1007/s11205-022-02943-4

Qasim M, Grimes A (2021) Sustainability and wellbeing: the dynamic relationship between subjective wellbeing and sustainability indicators. Environ Dev Econ 27(1):1–19

Rasoolimanesh SM, Jaafar M, Marzuki A, Abdullah S (2017) Tourist's perceptions of crowding at recreational sites: the case of the Perhentian Islands. Anatolia 28(1):41–51

Rasoolimanesh SM, Ramakrishna S, Hall CM, Esfandiar K, Seyfi S (2020) A systematic scoping review of sustainable tourism indicators in relation to the sustainable development goals. J Sustain Tour:1–21

Reinhold S, Laesser C, Beritelli P (2015) 2014 St. Gallen Consensus on destination management. J Destin Mark Manag 4(2):137–142

Riganti P, Nijkamp P (2008) Congestion in popular tourist areas: a multi-attribute experimental choice analysis of willingness-to-wait in Amsterdam. Tour Econ 14(1):25–44

Ritchie JRB, Crouch GI (2003) The competitive destination: a sustainable tourism perspective. Oxon, Cabi

Rosenow, J. E., & Pulsipher, G. L. (1979). Tourism the good, the bad, and the ugly.

Ruhanen L, Scott N, Ritchie B, Tkaczynski A (2010) Governance: a review and synthesis of the literature. Tour Rev 65(4):4–16

Ruhanen L, Weiler B, Moyle BD, McLennan CJ (2015) Trends and patterns in sustainable tourism research: a 25-year bibliometric analysis. J Sustain Tour 23(4):517–535

Schweinsberg S, Darcy S, Cheng M (2017) The agenda setting power of news media in framing the future role of tourism in protected areas. Tour Manag 62:241–252

Séraphin H, Sheeran P, Pilato M (2018) Over-tourism and the fall of Venice as a destination. J Destin Mark Manag 9:374–376

Séraphin H, Zaman M, Olver S, Bourliataux-Lajoinie S, Dosquet F (2019) Destination branding and overtourism. J Hosp Tour Manag 38(March):1–4

Séraphin H, Ivanov S, Dosquet F, Lajoinie B (2020) Archetypes of locals in destinations victim of overtourism. J Hosp Tour Manag 43:283–288

Shafiee S, Rajabzadeh Ghatari A, Hasanzadeh A, Jahanyan S (2019) Developing a model for sustainable smart tourism destinations: a systematic review. Tour Manag Perspect 31:287–300

Shi B, Zhao J, Chen PJ (2017) Exploring urban tourism crowding in Shanghai via crowdsourcing geospatial data. Curr Issue Tour 20(11):1186–1209

Sigala M, Hallen C (2019) The impact of social media on the behavior of wine tourists: a typology of power sources. In: Sigala M, Robinson RNS (eds) Management and marketing of wine tourism business: theory, practice, and cases. Springer International Publishing, pp 139–154

Sorokina E, Wang Y, Gyall A, Lugosi P, Torres E, Jung T (2022) Constructing a smart destination framework: a destination marketing organisation perspective. J Destin Mark Manag 23 (March):100688

Stewart WP, Cole DN (2001) Number of encounters and experience quality in grand canyon backcountry: consistently negative and weak relationships. J Leis Res 33(1):106–120

Stokols D (1972) A social-psychological model of human crowding phenomena. J Am Plan Assoc 38(2):72–83

Szromek AR, Kruczek Z, Walas B (2020) The attitude of tourist destination residents towards the effects of Overtourism–Krakow case study. Sustain 12(1):228

Teye V, Sirakaya E, Sönmez F, S. (2002) Residents' attitudes toward tourism development. Ann Tour Res 29(3):668–688

Trunfio M (2008) Governance turistica e sistemi turistici locali. Modelli teorici ed evidenze empiriche in Italia. Giappichelli Editore

Trunfio M, Campana S (2019) Drivers and emerging innovations in knowledge-based destinations: towards a research agenda. J Destin Mark Manag 14:100370

Trunfio M, Pasquinelli C (2021) Smart technologies in the Covid-19 crisis: managing touris flows and shaping visitors' behaviour. Euro J Tour Res 29:2910

Tyrrell TJ, Johnston RJ (2006) The economic impacts of tourism: a special issue. J Travel Res 45(3):3–7
UNWTO (2012) Global report on City tourism. In: AM reports: Volume six. UNWTO
UNWTO (2018) Overtourism? Understanding and managing urban tourism growth beyond perceptions
Verhoef PC, Reinartz WJ, Krafft M (2010) Customer engagement as a new perspective in customer management. J Serv Res 13(3):247–252. https://doi.org/10.1177/1094670510375461
Ward C, Berno T (2011) Beyond social exchange theory. Attitudes toward tourists. Ann Tour Res 38(4):1556–1569
Wymer W, Polonsky MJ (2015) The limitations and potentialities of green marketing. J Nonprofit Public Sect Mark 27(3):239–262
Yeh SS, Wai Aliana LM, Zhang FY (2012) Visitors' perception of theme park crowding and behavioral consequences. In: Advances in hospitality and leisure. Emerald Group Publishing Limited, pp 63–83
Zhang T, Chen Y, Wei M, Dai M (2022) How to promote residents' collaboration in destination governance: a framework of destination internal marketing. J Destin Mark Manag 24:100710

Chapter 4
Smart Technologies for Sustainable Tourism Development: Exploring Practices in European Destinations

4.1 Learning from the Practice: Smart Technologies for Sustainable Development

This chapter analyses selected cases concerning European destinations that have adopted smart technology tools for boosting sustainable development. Smart tools are often adopted occasionally and with a limited scope in international destinations (Ivars-Baidal et al. 2019). Often they cannot make a difference in pushing the destination onto a sustainable track (García-Hernández et al. 2019, p. 29). This case selection reveals tools and practices that will open up a discussion on advanced technology deployment, achieved specifically in a set of European destinations.

This practice-led research approach will cast light on different technological functions that, usually treated in the literature in relation to distinct theoretical constructs, combine digital multi-stakeholder engagement and smart destination experiences as drivers of sustainable tourism development. The adopted practice-oriented focus will feed a discussion contributing to integrative conceptual frameworks. It is largely unexplored in the literature and not mirrored by any integrative empirical perspective on the smart and sustainable destination (Eisenhardt 1989; Eisenhardt and Graebner 2007; Xiao and Smith 2006; Yin 2014).

The practice-led research effort aligns with the pyramid in Fig. 3.2, proposed as an analytical framework. The empirical research moves from the top of the pyramid regarding the adopted ICT tools and the related actions that introduce them as the visible layers of the underlying strategic framework and vision. These either pertain to a less accessible domain for researchers and observers or are not even defined a priori and/or shared by decision makers (see Chap. 3). Accordingly, the research design follows the proposed pyramid, as shown in Fig. 4.1.

The following sections will present different cases starting from the direct observation of tools and practices based on a review of secondary sources, including official documents shared by DMOs and local governments, official websites, online

C. Pasquinelli, M. Trunfio, *Sustainability-oriented Innovation in Smart Tourism*, Tourism on the Verge, https://doi.org/10.1007/978-3-031-33677-5_4

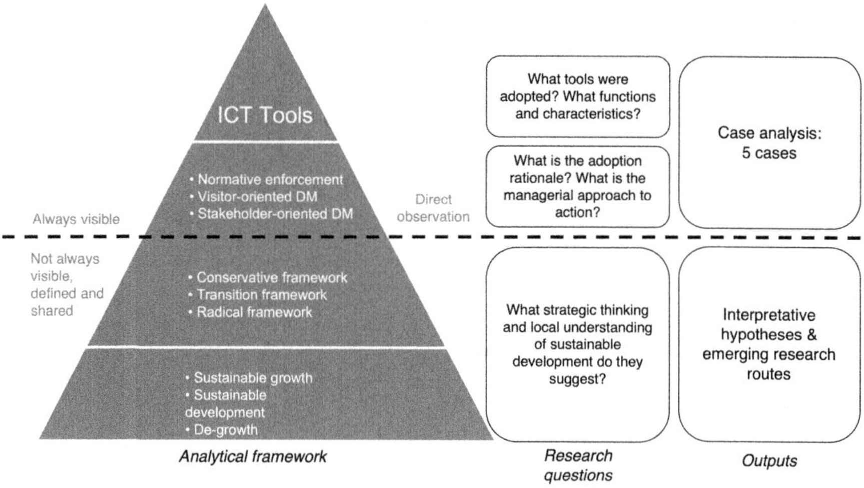

Fig. 4.1 From the analytical framework to the research design

news archives, and international public and private organisations' reports on smart and sustainable tourism.

This chapter will address the first two research questions in Fig. 4.1 (above the line of "visibility" and observation), whilst the concluding chapter will address the research question at the bottom of the diagram by discussing the results of the theoretical and empirical exploration carried out in this book.

The case analysis aims to provide fresh insights into how European destinations adopted and combined ICT tools that enhance smart tourists' experiences and engage destination stakeholders in the frame of emerging managerial approaches to sustainable development. It opens the route towards exploring the bottom layers of the built analytical framework (i.e. the pyramid in Fig. 3.2) and possible research hypotheses on sustainable-oriented innovation in smart tourism destinations. The cases were developed as illustrative cases of the exploratory potential of the proposed analytical framework and its utility in understanding policy mechanisms and visions that shape destinations' sustainable tourism development.

4.2 Smart Tourism Enhancing Experiences, Sustainability, and Accessibility: The Case of Genoa, Italy

4.2.1 The Genoa Tourism Destination

The metropolitan city of Genoa is the capital of the Liguria region in Italy. Liguria is the first region in Italy for the density of cultural heritage (270.9 assets every 100 km^2), and Genoa preserves theatres, museums, historical libraries, churches, and precious ancient buildings. Amongst Genoa's cultural heritage, the ancient Rolli Palaces, a monumental complex of 42 ancient buildings in the old town, became a UNESCO heritage site in 2006.

The city of Genoa presents a long tradition as a tourism destination thanks to the maritime economy and cruise tourism, combined with cultural tourism, events, and the food and wine experience. Several urban regeneration projects have restored the old commercial port, redesigned the waterfront (Aquarium, Congress Centre, Naval Museum) and renewed urban areas, revitalising economic and social places. Culture-led innovation allowed Genoa to be recognised as the European Capital of Culture in 2004.

In the last decade, Genoa has been growing as a tourism destination, and its cultural regeneration process, urban atmosphere, and coast have attracted international tourists. Genoa is one of the major Italian tourist ports, with more than 2 million cruise passengers and 2.5 million ferry passengers (2019). Urban and socio-cultural regeneration and the cruise industry's fast growth have generated thousands of tourists annually, challenging Genoa to invest in enhancing sustainable urban development conditions. Without a DMO, the Municipality of Genoa is in charge of tourism destination planning and marketing, promoting tourism development and sustainability, and facilitating innovative projects.

4.2.2 The Exploracity Immersive Experience

The *Exploracity* project, developed by ETT S.p.A., represents a pillar of technology-driven tourism innovation in Genoa.[1] It is an innovative platform that creates onsite smart experiences by exploiting diverse technologies such as a geographical information system, virtual and augmented reality, and gamification (Femenia-Serra et al. 2019; Neuhofer et al. 2012).

The platform transforms Genoa's immersive smart experiences, which combine the valorisation of physical attractions (churches and museums, forts and monumental gardens etc.) with immersive and virtual journeys into the past. It integrates the medieval city of Genoa's artistic and architectural beauties with immersive multimedia experiences, combining diverse multimedia contents (e.g. texts, images, and videos) and digital storytelling representations in two immersive itineraries, providing an unusual trip into history.

The first route, "Medieval Genoa", unfolds in the historic centre of Genoa through 15 points of interest, offering stories and anecdotes to discover the city's main attractions (Palazzo San Giorgio and Sottoripa, Piazza Banchi, San Lorenzo etc.) exploiting multimedia content. Various routes include historical information, images, virtual reality, and immersive technologies that create a new way of

[1] https://www.visitgenoa.it/en/exploracity-new-way-discover-city. Last access 9 December 2022.

knowing the city, live experiences of emotion, and escape and stimulate behavioural interactivity with digital content.

The second route promotes the "Gardens of Liguria", offering a VR experience based on text and photo content comprising eight points of interest that explore the landscapes, flowers, and plant nursery culture of Genoa and Liguria. Both itineraries are accessible by using a free app combined with a cardboard VR viewer (or alternatively with Gear VR), designed to ensure a device with a comfortable and easy-to-use level of usability and wearability to promote clever alternative access to destination information. The third and fourth journeys are being launched and concerned: The Churches of Genoa, an itinerary to discover 11 churches combining cultural and artistic content with religious narratives; and the Museum of Genoa, bringing visitors to 12 city museums through images and videos.

Exploracity allows a mixed reality experience that combines physical and dispersed visits (e.g. cultural heritage in the city) with immersive experiences (virtual travel). Several ICT tools have been combined (Pantile et al. 2018). The city's tourist information office offers an information touch monitor with photos and a video gallery showing the 15 main points of interest (POIs). Two virtual reality stations allow a virtual flight over the city of Genoa, during which POIs may be selected and virtually explored in 360-degree vision and a specific immersive experience of an exciting virtual walk, balanced on a high and narrow bridge within the Porta Soprana towers.

A mobile app, enriched with multimedia content in virtual and augmented reality, transforms the city journey into emotional and immersive experiences and drives post-experience behaviour (Trunfio et al. 2022). Exploiting virtual and augmented reality, Exploracity offers tourists the opportunity to immerse themselves in 360° videos or pictures and live the physical space (today's city experience) and the virtual space (the city experienced in the past). Diverse typologies of emotional and immersive experiences can enhance the value of the destination visit, such as education, heritage valorisation, escape, and entertainment (Trunfio et al. 2020). The cardboard VR viewer combines the real object (a building, church, monument, or square) and the superimposed virtual element (augmented reality reconstruction). Whilst listening to explanations and seeing various original monuments, churches, streets, and squares, all visible in their ancient magnificence, visitors can admire each secret corner of the city.

Gamification enhances the interactive and edutainment experience in the city, involving visitors, mainly younger tourists. Gamification represents the adoption of computer game technology in non-game contexts to enhance experience and engagement, bringing new hedonistic incentives, and encouraging and motivating further activities (Xu et al. 2013). The Ancient Crafts game, free and downloaded from the Exploracity app, guides visitors to the historical centre, offering general and historical information onsite and triggering virtual reality and augmented reality content whilst using the cardboard VR viewer. In addition, players can look for virtual objects representing ancient crafts, discover secret content, and gain points.

4.2.3 Smart Tourism Projects Enhancing Accessibility and Sustainability

Several smart tourism projects have been implemented in the city of Genoa. Co-financed through European funds, some of these projects were selected as leading examples of European smart tourism practices (Scholz and Agenda 2020), even though Genoa was not nominated as a European Capital of Smart Tourism. These projects addressed some issues at the core of the European Capital of Smart Tourism programme, such as digitalisation, sustainability, accessibility, cultural heritage, and creativity.[2]

The Municipality of Genoa invested in digital platforms creating digital accessibility during the pre-experience. The online interactive tourism map of Genoa allows people to plan and organise the city visit, defining specific tours and verifying the level of accessibility of each urban area.[3] The city's strategic vision builds on smartness and sustainability attempting to respond to both residents' and tourists' needs and expectations. The Municipality invested in facilitating tourism and entertainment for all, eliminating physical and informative barriers and creating an easy-to-use digital map for tourists and residents.

Genoa represents the first Italian city with a totally georeferenced PEBA (Plan for the Elimination of Architectural Barriers), with data free and accessible to all on a dedicated portal (open data) and a specific app. The PEBA maps the municipal area considering the perspective of motor and sensory accessibility and divides the city into 94 Areas, as portions of the territory where public functions of importance and interest for the citizen are concentrated.

Five thematic itineraries have been mapped with the level of accessibility, with main points of interest such as monuments, museums, universities, or historical shops. The itineraries are Art and Culture, Education, Free Time, Business, and Other Paths. In addition, the map offers the georeferencing of all forms of hospitality (hotels, bed and breakfasts, apartments etc.). Four colours identify the degree of accessibility considering the presence or absence of architectural barriers: green (accessible), yellow (partially accessible), orange (partially inaccessible), and red (inaccessible).

Metrominuto Genoa represents a complementary project where the most significant places have been represented as stops on various pedestrian lines connected to a network. It allows for designing public transport for pedestrian routes and the relationship with private vehicles, rethinking travel as a way of enhancing the city's beauty.[4]

---

[2] https://smart-tourism-capital.ec.europa.eu/index_en. Last access 9 December 2022.

[3] https://www.visitgenoa.it/en/mappa_interattiva. Last access 11 December 2022.

[4] https://www.visitgenoa.it/en/node/28795.Last access 11 December 2022.

Table 4.1 Genoa: Projects and ICT tools

Genoa projects	In brief	ICT tools
Exploracity	Onsite immersive smart experiences, which combine physical attractors' valorisation with immersive and virtual journeys in the past	Virtual and augmented reality Geographical information system Gamification Tourism information system
Online interactive tourism map of Genoa	Online accessibility (pre-experience) for planning and organising the city visit and verifying the level of accessibility of each urban area	Geographical information system
Metrominuto Genova	Designing public transport for pedestrian routes and the relationship with private vehicles	Geographical information system

4.2.4 Key Learning Points

The case of Genoa summarises a dynamic evolution of technology for smart tourism experience, stakeholder engagement, and sustainability (including accessibility and tourist flow dispersion). Three main projects have been implemented in Genoa, exploiting diverse and combined ICT tools (Table 4.1).

This practice shows how the destination is moving towards technology-driven innovations that leverage visitors' smart experience, and accessible and inclusive digital stakeholder engagement.

The city invests in a sustainable-oriented smart city that creates value for tourists and residents. Exploracity favours the smart tourist experience and disperses tourist flow in the city. Innovative projects exploit several smart technological tools for digital engagement and entertainment, enhancing the quality of life and tourism experience and moving towards sustainable and accessible tourism development.

The Exploracity implementation in the city of Genoa offers interesting contributions on how destinations invest in new realities to amplify the value of cultural heritage experiences. Smart technologies and stimulating digital engagement enhance the content of the cultural heritage exhibition (the physical environment) with additional multimedia representations of historical or current real life overcoming cultural heritage's spatial and temporal limitations. They provide new opportunities to virtualise and augment the traditional visitor learning experience in the museum, and visitors are immersed in unusual experiences whilst remaining in the familiar real world (Bekele et al. 2018; Schaper et al. 2018; Trunfio et al. 2020, 2022).

Exploracity and other smart innovations in Genoa have been capitalising on the value of smart technologies in the destination. They offer opportunities to reduce negative overcrowding effects by redirecting tourism flows from iconic sites towards new thematic routes (e.g. gardens) and stimulating changes in tourist behaviour.

4.4 Smart Mobility and Access to Reduce Congestion: The Case of Dubrovnik, Croatia

4.4.1 The Dubrovnik Tourism Destination

The city of Dubrovnik (Croatia) has a long and complex history. The Old Town is unique with its Gothic, Renaissance, and Baroque churches, monasteries, palaces, fountains, and squares and was included in the UNESCO World Heritage List in 1979. The tourism and hospitality industry represents the key pillar of the local economy. It can be traced back to classical antiquity when people visited Dubrovnik for relaxation, entertainment, and medical treatments.

Over the last few decades, the rapid growth of mass tourism has been driven mainly by cruise ship tourism (around 900,000 passengers annually), cultural tourism, and film tourism (e.g. Game of Thrones film locations). The monoculture of tourism generated high tourism impacts, exacerbating resident–tourist conflicts and urban and social imbalances and causing several adverse effects of overtourism, mainly in the Old Town (Panayiotopoulos and Pisano 2019).

Diverse unsustainable tourism practices and mass tourism marketing-oriented approaches have been adopted over the years, threatening the preservation of the Old Town cultural heritage and generating the fast degradation of residents' quality of life. Locals regularly protest against cruise ship tourism's negative impacts, similar to the case of Venice in Italy (González 2018; Capocchi et al. 2019). Unsustainable tourism behaviour and negative impacts on social life and the environment exacerbated the local community's reduced acceptance of tourism and the quality of the visitor experience (Trunfio and Pasquinelli 2021; Stecker and Hartmann 2019).

The incidence of number of tourists in the local population has increased quickly in the last decade. In 2016, Dubrovnik, with a population of around 43,000, received over 5 million tourists (about 19% of the Mediterranean market), with 10,388 cruise tourists in August alone. The World Heritage Committee of UNESCO observed that the city's World Heritage status was at risk, highlighting the critical and limited situation and calling for the sustainable management of cruise ships and determination of the maximum number of tourists in terms of the sustainable carrying capacity of the city.

The city launched a strategic tourism plan (focusing on cruises) and a city management plan to limit the number of tourists visiting the Old Town to 8000 in January 2017 to maintain UNESCO World Heritage status. They installed 116 CCTV cameras to count people entering or leaving.

The study "Sustainable cruise tourism development in Croatia", conducted by the Institute of Tourism, defined the number of cruise passengers, and the Dubrovnik Port Authority is in charge of implementing sustainable cruise tourism practices. The Dubrovnik Port Authority and the Dubrovnik County Port Authority monitor the limit of passenger numbers by a booking system monitoring the most congested days in a year and the number of cruise passengers.[11] An agreement defined the 2 + 1 principle: booking two large cruise ships at the port and one large cruise ship in front of the Old Town. The number of cruise tourists per day is 8000.

4.4.2 The Respect the City Project

The adoption of smart technologies to cope with overtourism imbalances that impact Dubrovnik's carrying capacity and the cultural heritage preservation of the Old Town can be dated back to 2015. In 2017, the Respect the City (RTC) project was launched as an innovative strategic destination management approach for Dubrovnik, which levered on various actions and tools, including ICT tools, to drive stakeholders towards sustainable and responsible tourism. The aims were to reduce the negative effects of overtourism, preserve cultural heritage, enhance the quality of citizens' daily lives, and ensure immersive and unusual experiences.

The project introduced several smart technologies. Destination management and marketing practices supported sustainable development, balancing culture, landscape and nature, and enhancing the visitor experience.

A *smart city action plan* was implemented to redirect tourism flows from iconic sites (Font and McCabe 2017), reduce overcrowding, and stimulate changes in tourist behaviour (Trunfio and Pasquinelli 2021), driving innovation in the Dubrovnik destination. Six key points constitute RTC's action plan: the definition of standard criteria and indicators to evaluate sustainable destination development; the increase of transparency in the management of environmental and natural resources by different destination stakeholders; the upgrade of the necessary infrastructure to manage crises and emergencies in the destination; reducing traffic congestion and improving traffic flows, innovating in researching new public transport; raising public awareness and increasing residents' satisfaction to involve their participation in City Planning; and the definition, measuring, and monitoring of indicators for small business and entrepreneurship, have been actively engaged in the destination's sustainable development.

Diverse technologies have been adopted and integrated into the smart city model, supporting the Respect the City project. They are innovative environmental management information systems composed of hardware, software, and mobile computing tools that collect environmental information to support destination managers' decision-making processes (Ali and Frew 2014a, b). The prevailing overtourism imbalance, which challenged the City of Dubrovnik, was the peak of tourists in the high season, mainly cruise ship tourists, which impacted the carrying capacity. The smart city project uses ICT technologies to measure tourists' presence and impacts—and then collects the data and insight into visitor movements and point of interest

[11] https://www.portdubrovnik.hr/dubrovnik-cruise-destination. Last access 11 December 2022.

preferences—providing information and redirecting visitor's flows (Trunfio and Pasquinelli 2021).

Dubrovnik launched the *visitor counting system*, comprising six city gates at the entrance to the Old Town centre and based on Artificial Intelligence which forecasts the number of visitors and cruise ship passengers, including the number of overnight stays, and their behaviour. Tourists can consult the number of visitors on the destination's website at each point of interest and decide where to go. In addition, an app gives users alternate routes and locations to visit when the number of tourists in the city exceeds 4000. Redirecting visitors out of the city represents a complementary purpose, which allows optimisation of the flows promoting alternative tourism offers.

The Rural Dubrovnik-Neretva web platform enables 84 destinations which propose rural tourism, wine and gastronomy tourism, and other thematic forms of tourism in the region north of Dubrovnik. Additionally, smart parking suggests how many parking spaces are available. The Dubrovnik Card typically provides city card advantages (discounts, unlimited use of public transport, and access to attractions) and via the iBeacons card, it suggests offers and other opportunities.

The city authorities can monitor the number of people in the Old Town in the City of Dubrovnik. They can make smarter and more informed decisions concerning the organisation of pedestrian movements around and inside the Old Town.

The integration between external and internal stakeholders has been facilitated by Dubrovnik Eye, an interactive web platform that enables direct communication between citizens and the Municipality. It points out and reports problems and issues related to tourism, including overtourism behaviour and forms of touristification (Trunfio and Pasquinelli 2021).

4.4.3 Smart Tourism Projects Enhancing Sustainability

The Integrated Action Plan for Dubrovnik as a Sustainable Tourism Destination defined a sustainable, smart, and socially conscious urban mobility system. This plan aims to improve citizens' awareness of smart mobility as a possible solution to unsustainable practices and introduces forms of traffic management supported by integrated smart solutions within public transport.

Smart micro-mobility projects underpin the sustainable development route undertaken by Dubrovnik City by exploiting smart technologies' power, enhancing sustainability and accessibility as leading issues of the European Capital of Smart Tourism (https://smart-tourism-capital.ec.europa.eu/index_en).

The Dubrovnik City City Changer Cargo Bike project capitalised on the best practices of several EU cities which implemented the EU pilot project City Changer Cargo Bike. They created a successful model for solving the problem of unloading cargo, especially in their pedestrian zones. The Changer Cargo Bike allowed the adoption of four electric cargo bikes in the City of Dubrovnik. They permit sustainable logistics operations that engage citizens (but also caterers, traders, delivery

Table 4.3 Dubrovnik: Projects and ICT tools

Dubrovnik projects	In brief	ICT tools
Respect the City (RTC)	Innovative strategic destination management approach driving stakeholders towards sustainable and responsible tourism	Destination Management system Environment management information systems Artificial Intelligence
Rural Dubrovnik-Neretva	Web platform enables 84 destinations which propose rural tourism, wine and gastronomy tourism and other thematic forms of tourism	Tourism Information System
Integrated multimodal public transport and parking	Smart micro-mobility projects, public electric marine transport and smart parking	Location-based services Tourism Information System

services etc.) in urban sustainability, reduce traffic congestion and pollution, and improve the quality of the public space in the historic centre.

Dubrovnik invests in an integrated multimodal public transport system that enhances sustainable local mobility. It developed public electric marine transport covering the city's coastal territory. Improving maritime communication increases public transport capacity and facilitates the dispersal of inbound and outbound visitors, reducing traffic and noise, and positively impacting sustainability.

By exploiting the smart technology opportunities, about 1900 sensors installed in Dubrovnik carried out one of the largest integrated smart parking solutions implemented on the Narrowband Internet of Things network worldwide. The application of free parking, available to citizens and tourists, allows people to find a parking spot online, reducing traffic jams, planning where it is possible to park, and saving time. Additionally, 20 public information displays have been implemented in some parts of the city to locate the free parking spaces close to the area.

4.4.4 Learning from the Practice

The case of Dubrovnik constitutes best practice on responsible and sustainable roles, practices, and behaviours, leveraging smart experiences and digital stakeholder engagement, with encounters between tourists and multiple local actors (local community and firms). Tourists' and residents' practices in Dubrovnik City enhance the quality of residents' life and the value of the tourism experience (Table 4.3).

The analysis underlines the crucial role of the sustainable tourism strategy, which exploits smart technologies to reduce tourist impacts and imbalances in Dubrovnik city. An integrated smart destination system leverages several smart technologies to drive decision-making and stakeholder behaviours in the destination (Ali and Frew 2014a, b; Trunfio and Pasquinelli 2021). Real-time technologies and big data address ordinary behaviour, dispersing flows and reducing tourism impacts.

Significant results have been achieved during the Respect the City project: reducing overcrowding in the old city centre; balancing the presence of cruise line tourists in the Dubrovnik port; improving smart and sustainable mobility; dispersing tourists outside the city, and enhancing new forms of experience in surrounding areas.

4.5 Smart Experience Design for the Metropolitan Ecosystem: The Case of Lyon, France

4.5.1 The Lyon Tourism Destination and the 2019 European Capital of Smart Tourism

Lyon is part of France's third-largest metropolitan area, with 1.4 million inhabitants. Tourism is an important economic engine, reaching 5.8 million overnight stays and over 34,000 jobs in 2021.[12] In the past, Lyon was mainly a business tourism destination for conferences and professional events. Today Lyon is also renowned for leisure and cultural tourism (Scholz and Agenda 2020) and attracts significant domestic/proximity tourist flows from the region (GrandLyon 2021). Since around 2010, this development has been systematically triggered by the aim to rebalance inflows over the week and seasons. From 2011 to 2019, the hospitality industry in the city recorded a + 20% occupation rate during the weekends, witnessing the expansion of leisure tourism (GrandLyon 2021). Business tourism continues to be a relevant component of the industry, with over 100 annual events, including congresses, trade shows, fairs, conventions, and meetings.

The widely renowned OnlyLyon brand assisted tourism development and the city-region's positioning as a leisure and urban tourism destination (GrandLyon 2021). Lyon's branding results from a collaborative scheme involving local stakeholders in the metropolitan area, framing international marketing actions and overcoming historical fragmentation and lack of coordination (Eurocities 2010). The OnlyLyon brands' success resulted in clear leadership through decision-making and the active engagement of the partners, who agreed to rely on the city brand to disseminate their international and national communication.

The Lyon city centre is listed on the UNESCO heritage list; it is the gastronomic capital of France, and the "Bouchons Lyonnais" quality label assures a sustainable local approach to the traditional gastronomic offering in the city (Scholz and Agenda 2020). The city hosts the internationally renowned Festival of Lights, which greatly contributes to the brand image.

The city has placed sustainability at the core of urban development. The label "Lyon, Ville Equitable et Durable" was adopted by around 200 firms, events, and

[12] https://presse.lyon-france.com/en/tool-box/key-figures#:~:text=Figures%2C%20stats%20and%20analyses,and%2034%25%20leisure%20tourism. Last access 2 December 2022.

retailers to identify themselves as underpinning responsible behaviour and consumption and making significant investments in accessibility. In 2017 the city achieved the Access City Award.[13] Then, in 2019, building on a long-standing path of investments in smart and sustainable development, Lyon was awarded the title of European Capital of Smart Tourism. The EU programme rewards smart tourism practices and innovative tourism in European urban contexts, focusing on accessibility, sustainability, digitalisation, cultural heritage, and creativity.

Today, national tourism represents 75% and European tourism 12% of incoming tourists; the top foreign regions of origin are proximate markets in Europe and the USA (GrandLyon 2021). The sustainable tourism plan for the Métropole de Lyon (*Schéma de Dévelopment du Tourisme Responsible 2021–2026*) emphasises the relevance of targeting local tourists coming from the territory, with a potential of 1.3 million local tourists across the year, through dedicated communication tools and strategy and further promoting the Lyon City Card 365, designed for residents in the metropolitan area (comprising 59 towns).

Since 2010, tourism development and planning have become the competence of the metropolitan authority, thus positioning tourism in the inter-municipal frame for planning (GrandLyon 2021). The current sustainable tourism plan capitalised on and furthered the sustainability strategy of the Lyon metropolitan region, aiming "to make Lyon a benchmark destination for sustainable tourism".[14] The need to "act collectively to transform the city's approach to tourism", the need for more balanced tourism development, creating value for the territory and creating jobs without boosting residents' rejection, frame the sustainability vision of the Métropole de Lyon (GrandLyon 2021). According to the plan, all stakeholder engagement with responsible tourism is instrumental to sustainable tourism development. These are the key objectives stated in the plan: (1) consolidating the achieved balances through past strategies implementation, (2) boosting employment in the tourism industry, (3) acceleration of the environmentally friendly offering, and (4) inclusive tourism.

Tourism is integral to the city-region's sustainable development. The tourism development plan was developed by the Tourism division in the Métropole de Lyon's organisational chart, under the division in charge of territorial promotion and international relations. It is part of the *Délégation Développement Responsable,* which is the unit in charge of sustainability policies at the metropolitan level. That is, tourism development is tightly and unequivocally coupled with sustainable development. This seems to suggest that there is no tourism development policy outside of sustainability considerations for the region.

Multi-sector, multi-stakeholder, and inter-municipal governance implement the metropolitan tourism development strategy. A Destination Council (*Conseil de*

---

[13] https://smart-tourism-capital.ec.europa.eu/lyon-winner-2019-european-capitals-smart-tourism_en. Last access 3 December 2022.

[14] https://business.onlylyon.com/en/discover-lyon/business-sectors/sustainable-tourism. Last access 2 December 2022.

destination) was designed to follow the development of the actions and assess them, monitoring and adjusting the strategy over the next 5 years (GrandLyon 2021).

4.5.2 The OnlyLyon Ecosystem: A Long-Term City Development Path

Through a long-standing development path, recently culminating in the European Capital of Smart Tourism award, the city of Lyon and the metropolitan area are progressively configuring as the "OnlyLyon ecosystem". This is how the website OnlyLyon frames the overall strategy of city management and marketing ("A website of the OnlyLyon ecosystem"). The website is the umbrella platform encompassing *OnlyLyon* as the city brand. It focuses on the locals as city ambassadors, *Visit Lyon* for visitors, *Your Business in Lyon* for investment promotion, *Your Convention In Lyon* for convention bureaux activities, and *Study in Lyon*, which targets students for the local education cluster.

These priorities were supported by the adoption and integration of several ICT solutions, such as the Lyon City Card, the central booking portal, digital support for accessibility for visitors with disabilities, enabling an autonomous experience of the city, innovative signage in the historical sites of Lyon, and an augmented reality and geolocalisation app. This app helps visualise the evolution of city areas where tours are offered, following a path of painted murals signalling UNESCO-listed sites in other more peripheral districts (Scholz and Agenda 2020).

The Lyon City Card is the urban pass that includes museums, attractions, public transport networks (including self-service bikes), and guided tours. It gives discounts and priority access and is all-inclusive and user-friendly, achieving a high satisfaction rate (98%, Scholz and Agenda 2020). Two features are particularly relevant for supporting the sustainability goals: on the one hand, it gives visibility to off-the-beaten-track districts and lesser-known museums, contributing to dispersing tourists; on the other hand, particular attention is devoted to residents, designed for the needs of residents and visitors at the same time (Scholz and Agenda 2020).

The innovative pedestrian signage in Lyon's historic sites aims to guide visitors through the UNESCO sites in the city and avoid overcrowding. Street kiosks, direction signposts, orientation tables equipped with the NFC (Near Field Communication) system, and flash codes for app access comprise a digital signage network with real-time personalisation of the information and inputs provided.

4.5.3 *The OnlyLyon Experience: Towards Big Data-Based Experience Design*

In 2019, the development of the smart tourism capital continued with the introduction of the OnlyLyon Experience, an ICT tool that further assists the sustainable tourism destination. The OnlyLyon Experience is a destination CRM system involving public and private partners in collecting and sharing data and insights. Hotels, museums, and service providers feed the central database, which allows knowing more about visitors, their preferences and their needs. All key tourism stakeholders can rely on data for customising and tailoring the communication and experience offering. Omni-channel and omni-partner marketing actions are enabled by this tool through an effort of collective marketing, building with the partners a "communication architecture" for addressing different visitor groups.[15]

The goal is to create a "unique customer relationship", "an excellent and memorable experience in Lyon", and an "emotional, lasting and profitable relationship with local, national, and international visitors".[16] The messages sent to visitors are meant to improve customers' experience. The system also provides itinerary suggestions that may reduce congestion and traffic in crowded areas/times of the year (Scholz and Agenda 2020).

Equal attention is paid to locals and national/international visitors, all considered city users and visitors. This ICT tool is designed to support visitors from the pre-visit, throughout the visit, and in the post-visit. The system integrated the OnlyLyon Tourism customer databases and the partners' databases (e.g. AccorHotels Group, Hôtel Axotel, Lyon City Tour, Les Bateaux Lyonnais, Musée des Confluences, and the Hard Rock Café). Data collection is coupled with OnlyLyon Tourism's offering of operational services and customer knowledge-sharing activities to assist all the partners towards exploiting the system's potential and the high-quality design of experiences and digital communication.

There is an issue of privacy and data collection and retention whose responsibility is on OnlyLyon Tourisme et Congrès, who lead the system. Data collection and the processing of personal data are motivated by client and prospect management, as allowed by the law. As the Privacy and Confidentiality Policy states,[17] data processing goals include contracts, orders, invoice management, customer relationship monitoring, the definition of customer loyalty programmes, product testing and promotion, and the development of sales and client knowledge statistics. Collected data include (here a selection): identity (title, name, address, date of birth); payment methods; details of the purchase; family, economic and financial situation such as marital status, number of people in the household, number/age of children, profession, presence of pets; information about the business relationship (purchases,

[15] https://smart-tourism-capital.ec.europa.eu/lyon-winner-2019-european-capitals-smart-tourism_en. Last access 2 December 2022.

[16] https://en.lyon-france.com/onlylyon-experience. Last access 3 December 2022.

[17] https://en.lyon-france.com/terms-of-sale#privacy. Last access 3 December 2022.

Table 4.4 Lyon: Projects and ICT tools

OnlyLyon projects	In brief	ICT tools
OnlyLyon Web Portal	"A website of the OnlyLyon ecosystem" Umbrella platform (OnlyLyon, Visit Lyon, Business in Lyon, Your Convention in Lyon, Study in Lyon) Central booking	Destination management system
Lyon City Card	City users/tourist card (visit planning, transport networks, booking, discounts, priority access)	Destination management system and tourist/user card
Digital pedestrian signage (e.g. street kiosk, NFC system, orientation tables, and flash codes)	Guiding visitors in UNESCO sites and minor attractions, providing real-time personalised information	Location-based services
AR/geolocalisation app	Experiencing the evolution of city areas throughout the painted murals itineraries	Location-based services Augmented reality
OnlyLyon Experience	Destination CRM system for experience personalisation and data repository for destination knowledge production	Tourism Information System Destination management system

delivery address, frequency, comments from clients etc.); information on the type of promotional offering proposed to clients; data related to clients' engagement, sharing opinions on products and services and contents.

This wealth of data (in 2020, the database had about 2 million contacts, Scholz and Agenda 2020) provides the opportunity to build a knowledge system. Data collection per se is not enough to produce actionable knowledge and actionable knowledge for all the partners. So, it is important to comment on the operational services and knowledge development and sharing activities carried out by OnlyLyon Tourism. One of the challenges of destination management is micro-data development for planning and decision-making: the OnlyLyon Experience is a double-headed tool in this regard. It offers visitors better and more personalised experiences and offers tourism organisations knowledge for designing high added-value actions.

4.5.4 Key Learning Points

The case of Lyon has to be interpreted in the frame of a long-term trajectory of sustainable development. Table 4.4 summarises the described technology-based projects discussed in the previous sections.

This trajectory is characterised by constructing a city-region planning framework for tourism development and, on the other side, the expansion of tourism from business to leisure tourism. The tourism performance of the city and the metropolitan area proves the strategies' success. The European Capital of Smart Tourism title showed the destination's consolidated path, signalling the innovative approach to the sustainable destination that has developed over time. In 2021, Lyon recorded the best improvements according to the Global Destination Sustainability Index and ranked second as the best European destination (President of OnlyLyon Tourisme et Congrès, LinkedIn post, 2021). The current sustainable tourism plan for the Métropole de Lyon declared its ambition to be a global destination benchmark for sustainability. What is remarkable is the strong link between tourism planning and sustainable development, mirrored by tourism governance.

Lyon's development trajectory has seen constant technological investments, showing progressive advancements and sophistication. The OnlyLyon Experience is the most recent project described in this section and is remarkable for arriving after several ICT tools deployed over time. It suggests many strategic directions: first, customer relationship building in the pre-, during, and post-visit, working on customer loyalty, and repeat visitors. The relevance of local and national markets suggests the value of this direction. Second, the offering of relevant technologies not only to tourists but also to residents with different needs and perspectives, so integrating typical visitor functions with those relevant to people living in the city or metropolitan area. Third, data production and knowledge development for destination planning, management, and marketing.

This aspect suggests how this technological evolution builds on the long-standing practice of networking and collaboration amongst public and private stakeholders, developing trust and commitment to make this project realistic and work. That is, this case suggests a matter of timing in technology adoption for the destination as a system, which influences the technological investment's feasibility and probability of success.

4.6 Building the Sustainable Tourism Knowledge Ecosystem: The Case of Goteborg, Sweden

4.6.1 The Goteborg Tourism Destination and the 2020 European Capital of Smart Tourism

Goteborg is a destination attracting over 3 million tourists in 2021 (−35% compared to 2019). The Swedish market is the most relevant, representing 70% of visitors, whilst 30% of incoming visitors are foreigners (Goteborg and Co 2021). Business tourism related to professional meetings and events is a significant component of urban tourism. After the pandemic shrank this sector, many efforts were made to consider the future development of business tourism in the region. Cruise tourism is another component of local tourism, with 68,000 passengers arriving at the port of Goteborg in 2021, mainly from Germany (Goteborg and Co 2021).

Destination performance is monitored and reported annually in the Annual Sustainability Report, showing a 2021 visitor index high, measuring visitors' feeling of safety, attractiveness, availability of information, and perceived treatment during their stay (80, the target by 2030 is 84). 87% of residents in a survey sample reported that the visitor industry adds value to them (the target by 2030 is 90%). Since 2016, Goteborg has confirmed its top position in the Global Destination Sustainability Index, the first position confirmed in 2022.[18]

Goteborg & Co, the destination company part of Goteborg's Stashus AB that manages the companies owned by the city of Goteborg, is committed to a sustainable city and to offer "sustainable, attractive and unique experiences".[19] Goteborg & Co manages and markets the destination, comprising 13 municipalities (Goteborg and Co 2021). Over 30 years of activities, Goteborg & Co has built its centrality in the region and is the pivot of regional collaborations in the tourism industry, connecting public and private actors and academia (Masuch and Norman 2021, unpublished).

Its mission is to confirm Goteborg as a global example of a sustainable destination, to support and encourage local tourism organisations to undertake a sustainable trajectory and to "inspire and help" visitors to experience the city-region sustainably. Goteborg & Co's strategy aims to strengthen the metropolitan region as an inclusive and accessible destination. The aim is to maintain Goteborg as a growing destination in terms of job creation, economic growth, and addressing social issues. Efforts towards reducing the environmental impacts of tourism activities are made towards an "environmentally smart and climate-smart destination".

In 2020, Goteborg was nominated European Capital of Smart Tourism by the European Commission, acknowledging the long-standing and progressive smart approach to sustainability. Key projects concerned accessibility for all, for instance, redesigning the Liseberg Amusement Park and designing an accessibility app for Got Events, dedicated to people with disabilities. Goteborg boosted the adoption of environmentally friendly labels in the tourism industry, reaching 95% of the city's hotel rooms with an environmental certification (Scholz and Agenda 2020). Actions for spreading visitor flows included the "Next to Goteborg" campaign facilitating island visits by improving physical and digital accessibility. Significant digitalisation investments were made, and the pandemic accelerated this process, with the awareness that new expectations and preferences will impact the industry and the overall destination (Goteborg and Co 2021).

4.6.2 The Virtual Destination for the Sustainable Experience Industry

Different projects feed the virtual destination as a platform for analysing, planning, and monitoring. The Digital Twin of Goteborg visualises and simulates urban development projects. It visualises data of different types in a model mirroring the

[18] https://goteborgco.se/en/about-goteborgco/sustainability/. Last access 12 December 2022.

[19] https://goteborgco.se/en/about-goteborgco/sustainability/. Last access 12 December 2022.

whole surface of the city (about 700 square kilometres). Led by the urban planning department, it is relevant for all departments at the Goteborg Council. It is a 3-D copy of the city with all the architectural and natural features and details. It helps to understand how the city functions, have real-time control over it, develop scenario analysis, anticipate events, and take more informed decisions based on viable simulations.[20] Adopting the digital twin responds to the need to make better decisions faced with rising complexity that is determined by climate change and social imbalances.

Virtual Goteborg is meant to support the development of valuable services for residents, visitors, and companies and contribute to harmonious city development. It is a platform making the spread of information easier and enables effective participation and transparency.[21]

Through the support of the digital twin, Xperience Next promoted a full-scale experiment project for destination development regarding the island of Kållandsö.[22] Xperience Next is a programme at the Lindholmen Science Park that promotes innovation and is a catalyser for change. It runs various projects, including the acceleration of the "sustainable experience industry" through digitalisation, business development, and technology.[23] The full-scale experiment in Kållandsö consisted of observing the destination carrying capacity through a gaming tool that loads consumption data from local businesses, connecting with the digital twin to create simulations. Entrepreneurs can join the game by experimenting with different choices and behaviour to analyse how the group of gamers' decisions impact the destination and to what extent an equilibrium is achieved. The gaming combined with the simulation tool helps to assess, for instance, what happens if the parking areas are enlarged or the number of rentals increases.

In addition to the virtual destination projects, in 2021, the Hybrid+ project was launched to create better digital experiences by hybridising the physical experience at the destination. This project was launched in the frame of the actions revamping professional events and meetings sector, highly impacted by the pandemic. The goal is to innovate the business event destination, providing cutting-edge solutions for future evolutions in hybrid events (Goteborg and Co 2021).

[20] https://stadsutveckling.goteborg.se/digitaltvilling/. Last access 3 December 2022.

[21] https://stadsutveckling.goteborg.se/sa-planeras-staden/digitaltvilling/nyheter/pris-till-tvillingen/. Last access 3 December 2022.

[22] https://xperiencenext.lindholmen.se/sv/projekt/building-investment-lab-digital-twin. Last access 12 December 2022.

[23] https://www.lindholmen.se/en. Last access 3 December 2022.

4.6.3 Destination Data Platform: Towards a Tourism Knowledge Ecosystem

In 2020, Goteborg & Co initiated the development of a Destination Data Platform (DDP). This ambitious project progresses from the awareness that knowing visitors and their behaviour is the key asset for competitive destinations and tourism organisations (Masuch and Norman 2021, unpublished). The aim is to create a visitor industry's knowledge hub that helps organisations learn about visitors (what people think and do before, during, and post-visit), interact with them through real-time data, and predict future trends.[24]

The DDP collects data from several suppliers, which may be willing to distribute their data as open data, and generates insights as a service supporting different users. The partners provide data about transport, accommodation, tours, activities, and shopping (e.g. credit card companies, telecommunications, and transport hubs) and are intersected with data from local businesses. Micro-data statistics, forecasts, and real-time monitoring are created (Masuch and Norman 2021, unpublished), which may assist visitor flow control and management. For instance, data from credit card companies, which inform about consumption and purchase patterns, and data from mobile phone companies providing information on movements in the city, can be combined with data provided by the local hospitality industry, giving a detailed picture of the actual behaviour of visitors.

Building on its relational capital built over decades, Goteborg & Co is developing the network of actors that feed the knowledge hub and, using it, is creating the knowledge ecosystem that underlies the DDP. Hospitality businesses, visitor industry players, business analytics companies, booking platforms (which may buy datasets to be processed autonomously), and other destinations may benefit from the DDP. The existence and easy access to the knowledge ecosystem may also facilitate collaborations and partnerships amongst tourism organisations and amongst destinations that join the platform, thus improving local tourism innovativeness (Masuch and Norman 2021, unpublished).

The DDP is designed to directly support data-driven decision-making for marketing plans and business development actions. Local tourism organisations may be willing to provide their data to have data analyses and insights into visitor behaviour ready to be processed in decision-making and marketing plans. This allows tourism businesses to go beyond raw indicators such as overnights and occupancy rates and understand what the visitors prefer, do, and buy when visiting the city.

In line with the DDP approach, the Tourism Agenda project in the frame of Xperience Next programmes aimed to develop a big data-based tool to forecast trends and anticipate visitors' needs. This is functional to the development of sustainable experiences. Business cases were developed by Goteborg & Co

[24] https://goteborgco.se/en/our-services/knowledege-and-statistics/the-knowledge-hub/. Last access 3 December 2022.

Table 4.5 Goteborg: Projects and ICT tools

Goteborg projects	In brief	ICT tools
Accessibility app for Got Event	Enabling and improving access for people with disabilities	Location-based services
Virtual Goteborg	Digital twin	Computer simulation Geographical information system Virtual reality
Kållandsö project	Full-scale experiment through gaming and integration with the digital twin	Carbon calculator Economic impact analysis software
Hybrid +	Hybrid (physical and digital) experiences for the professional meeting sector	Virtual /AR reality
Destination Data Platform	Knowledge hub, collecting and processing data and offering analytics and visitor insights, for data-driven decisions	Tourism Information System
Event Impact Calculator	Forecasting the economic, social, and environmental impacts of urban events	Carbon Calculator Economic impact analysis software

concerning local events, such as sports events and concerts.[25] Concerning events, the city of Goteburg relies on the Event Impact Calculator, a forecasting tool developed in the city that helps event organisers and local authorities assess the economic, social, and environmental impact of the event (Scholz and Agenda 2020).

4.6.4 Key Learning Points

The case of Goteborg has shown the importance of the leading role of a sustainability vision which emerged over decades, followed by and associated with constant technological advancements. Table 4.5 summarises and categorises the ICT tools in Goteborg, presented in the sections above.

The sustainability framework identifies needs and priorities that the deployment of ICT tools may help to meet. Goteborg & Co, the official destination management organisation, not only invests in the acquisition of existing technologies but also plays a leading role in the design and implementation of tourism-dedicated technologies.

Developing and adopting the ICT tools described reveal the importance of collaborative capacity, networking, and relational capital. The case of DDP development is an example. Goteborg & Co, which defines itself as a platform for

[25] https://xperiencenext.lindholmen.se/sv/projekt/measuring-economic-values-experience-industry-through-big-data. Last access 3 December 2022.

collaboration,[26] is an active relationship builder with local, national, and international partners. The local entrepreneurial context is traditionally keen on collaboration and partnerships to improve the visitor experience (Masuch and Norman 2021, unpublished). Technology, data management, and science challenges may build on this characteristic of the local tourism system. The maturity of the local relational system can be the necessary starting point towards building a knowledge ecosystem. Knowledge is a critical asset, and the perceived value and safety in sharing knowledge is a prerequisite for the technological advancement pursued in the case of Goteborg.

In Goteborg, most ICT projects discussed in the previous sections emphasised the relevance of B2B smart technologies. These tools aim to build and structure the tourism business and destination offering and their marketing approaches rather than focusing on direct usability by visitors. Although the final target is the visitor, most of the presented projects are meant to inform and guide public and private stakeholder decision makers through data analysis and simulations.

4.7 Exploiting Smart Experiences and Digital Stakeholder Engagement for Sustainable Destination Development

The European cases have provided insights into adopting and exploiting ICT tools and their combinations to pursue sustainable destination development. Diverse usage rationales have emerged, including destination experience communication and marketing, aimed at shaping visitors' onsite experiences (targeting the visitor, either being a tourist, coming from outside the city or region, or a local destination user); the collection and combination of data to support monitoring, forecasting, and decision-making; and stakeholder relations building towards boosting better and sustainable destination experiences.

Overall, it is remarkable how in none of the analysed cases, investments in technologies are aimed to exclusively support centralised decision-making and actions by the DMO or local governments. All the described projects and tools were designed to assist dispersed decision-making, planning, grassroots actions, and widespread responsible behaviour. These tools were designed, implemented, financed by local authorities, and managed by DMOs. Still, they were meant to assist the different stakeholders' decision-making, from the DMO's decisions and action planning, to the individual tourism organisations planning their offering, and the individual tourist's choices and behaviour. These cases highlight that the sustainable destination is necessarily built on multi-stakeholder engagement and actions.

Part of the described projects were dedicated to the local tourism system, hence to tourism stakeholders in a narrow sense; on the other hand, the majority of projects adopted an integrated logic since they encroached on tourism industry borders and

[26] https://goteborgco.se/en/about-goteborgco/. Last access 11 December 2022.

involved residents, workers, and entrepreneurs only indirectly connected with the tourism sectors. The case analysis provided evidence of how ICT-based projects can be designed to respond and contribute directly to the broader smart city vision (Errichiello and Micera 2017; Camero and Alba 2019; Cavalheiro et al. 2019). That is, the analysed cases highlighted that the sustainable destination is necessarily built on the engagement and actions of multiple urban stakeholders beyond their direct "stake" in tourism.

Adopting ICT tools oriented to sustainability has emerged as a threefold approach. First, it contributes to planning and managing destination products, services, and visitor flows; second, through enabling and digitally enriching the destination experience; third, motivating and, at the same time, leveraging multi-stakeholders' engagement. These approaches, conceptualised and sustained in the literature, lacked empirical validation. The presented case analysis has provided preliminary explorative insights into the analysed European practices.

The case analysis drew attention to the opportunity to conceive smart experiences and stakeholder engagement as two complementary faces of the process of technology-driven innovation for sustainable development. That is, smart experiences engage visitors, make them more likely to engage with the place and the local community and share data and insights. Digital tourism stakeholder engagement assists the coordination and collaboration for better, smarter, and more sustainable destination experiences. It is worth noticing the role of open data production, collection, and analysis as an essential glue for stakeholder engagement and relevant smart experiences.

4.7.1 Smart Experience for Sustainable Development

Advances in technological devices allow an impact on a visitor's sensorial, affective, cognitive, relational, and behavioural responses (Lemon and Verhoef 2016; Homburg et al. 2017; Trunfio et al. 2022). The literature on smart experience for sustainable development is marginal (Trunfio and Pasquinelli 2021). Interpreting our findings with a new lens can contribute to the academic debate on sustainability-oriented smart innovation.

The ICT projects implemented in the European destination can be explored by considering their role in reducing tourism impacts and overtourism imbalances and redirecting stakeholders' behaviour. Smart experiences allow visitors' behaviour to be addressed and managed. Exploiting big data for tourism behaviour analysis and profiling visitors and, then, for proposing tailor-made experiences; monitoring tourism flows and forecasting their intensity over time, visit intentions, and preferences. Smart experiences can also be conceived as virtual remote experiences, which may partially replace the onsite visit, anticipate it, and enrich it with content and an "expert attitude" to the destination.

The dispersal and redirection of the tourist flow represent the first effect of the smart experience on enhancing sustainable development. Exploracity in Genoa provided an example of an onsite smart experience, hybridising digital and physical elements that boost an immersive experience. It combines a physical visit to the city's cultural heritage enriched by several virtual and augmented immersive experiences that drive visitors towards several itineraries. Gamification amplifies the edutainment and supports flow delocalisation.

In the Dubrovnik destination, the Respect the City project redirects tourism flows from iconic sites, reduces overcrowding, and stimulates changes in tourist behaviour. In the case of Lyon, an augmented reality and geolocalisation app pursue the same goals. Both tools channel visitors into a set of itineraries whose design combines the value of the visitor experience with visitor flow management objectives. In the case of Lyon, the city/tourist card, addressing both tourists and residents, facilitates access by adding information, booking, pricing, and discount options that support the visitor's construction of the experience. The smart experience has emerged as a modality to pursue a sustainable destination. Through a promise of value co-creation to the visitor, it may successfully persuade and convince visitors to direct their attention and discovery efforts towards off-the-beaten-track and lesser-known sites and itineraries, which may contribute to rebalancing the social, economic, and spatial impacts on the destination.

The analysed cases provided evidence of how the smart experience is enabled by social and technological factors, opening up space for experience co-production (Buhalis and O'Connor 2005; Hjalager 2015; Lam-González et al. 2019). It is remarkable how the smart experience allows visitors to decide where to go and what to do in the destination, thanks to tools that provide real-time and constant information on crowding, availability of space, and the opportunity to have a quality visit in a certain destination area. Examples include the visitor counting systems and smart parking solutions in Dubrovnik and the digital signage in Lyon's historic sites, integrating the NFC system for personalised and real-time onsite information.

To be noted is the role of the smart experience as a frame to connect the metropolitan area, which is central in Lyon and Goteborg. The smart experience can represent a modality to connect the urban centre to the region's cultural offering, pushing tourism's social and economic impacts towards additional areas needing tourism development. The smart experience can help visitors to pre-taste the value of moving outside the tourist hotspots and to visualise the opportunity to enrich their urban experience with additional valuable experiential elements.

There is an evident shift of attention towards smart experiences for all, acknowledging that smart experiences, as enabled and enhanced by technology, are the only possible quality, independent, and authentic experiences for people with disabilities. The presented case analysis reports broadly about the accessible experience for all (e.g. Genoa, Lyon and Goteborg).

A relevant point has emerged in relation to visitor engagement with the smart experience, encouraging visitors to get involved with the smart destination experience over the visit, being open and likely to value the education, heritage valorisation, escape, and entertainment content of the smart experience (Trunfio et al. 2020). The tourist system can enable the smart experience, but the willingness to try it and get involved depends on the tourist and his or her propensity to engage with it. In literature, gamification was discussed in this regard (Buhalis et al. 2019; Buonincontri and Micera 2016; Neuhofer et al. 2012), and the case of the Faroe Islands gives some insights into virtual tourism experiences. The possibility to remotely control locals' movements onsite created an element of curiosity and interactivity typical of gaming experiences.

The case of the Faroe Islands drew attention to another aspect of visitor engagement which have remained largely unexplored in the literature. This concerns the adoption of tailor-made technology that emphasises the unique destination brand message: in the case of the Faroe Islands, this was the contemporary and technology-savvy way to be remote. The smart experience should be considered the "local" experiential narration of smartness, as a way to state how the smart challenge is addressed and what it means to the specific destination and its community.

The case of Lyon with the OnlyLyon Experience, an advanced destination CRM system, draws attention to the smart experience as it is constructed, thanks to smart devices that, partially visible to the tourist, are rooted in the destination network. That is, beyond the smart experience as physical, virtual, or hybrid experience enabled and enhanced through technology, there is an important aspect to consider about the smart building of the tourist experience. The experience is "smart" as articulated through the CRM system, to be personalised and functional to create value for visitors and the destination. In addition, the smart experience also configures as producing digital content and big data emerging from visitors' choices and behaviour. This opens up the role of smart experiences in motivating the destination stakeholder's engagement, which is discussed in the following section.

4.7.2 Digital Stakeholder Engagement for Sustainable Development

The analysed cases have provided insights into how ICT infrastructure stimulates and is meant to enhance stakeholder engagement (Cabiddu et al. 2014; Sigala 2018; Trunfio and Della Lucia 2019). The case of Lyon and Goteborg has cast light on this.

Experience building and product management motivate local stakeholders to join and actively contribute to the OnlyLyon Experience and Virtual Goteborg. In the case of the OnlyLyon Experience, the system is double-headed, having direct relevance and visibility to visitors and tourism organisations. In the case of Goteborg, Virtual Goteborg, and the Destination Data Platform are, at the moment, business-to-business smart tools which create the premise for better and smarter visitor experiences.

These ICT solutions are fed by the individual stakeholders' raw data and provide analytical insights allowing businesses and tourism organisations to focus on business development and planning. Technology is not only the enabler of stakeholder engagement but also the motivation for becoming interested in active participation, which requires stakeholders to "donate" an important asset (i.e. data) to the local system and to get back value (i.e. systemic knowledge). In the two cases, the DMO carries the costs and responsibility for the acquisition of micro-data (also in terms of respect and a guarantee faced with national and international laws on data protection), something that would remain a burden on the single stakeholder. This burden is unrealistic for many tourism players. The acquisition and analysis of micro-data in tourism systems, which is so important and difficult for the limitations of the available official statistics, can become a glue for the destination network.

Virtual Goteborg draws attention to the experiential learning that the digital twin may provide for tourism organisations. In particular, the Kållandsö project integrated gaming mechanisms to support tourism operators in learning about the impact of their actions and decisions through simulations and an impact calculator. The smart destination experience may be conceptually opened up to involve destination stakeholders other than visitors, as the emerging ICT tools go in this direction. The opportunity for effective learning, particularly about sustainability management, may represent another motivation for becoming interested in active participation. In this case, the local stakeholder needs to give time and be willing to learn whilst receiving in return a learning opportunity and analytical insights in terms of forecasts and simulations that potentially inform transitions towards more sustainable business models.

The case analysis has discussed tools and practices focused on internal stakeholder engagement. A willingness to be engaged is steered by providing concrete opportunities for value co-creation in the local system, also creating opportunities for collaborations amongst the stakeholders for collective marketing efforts and tourist product development (as has emerged in the case descriptions). Less visible was the aspect emerging from the existing literature about forms of e-democracy (Sigala and Marinidis 2012), framing the role of ICTs that empower stakeholders (business stakeholders and residents) to actively participate in bottom-up decision-making processes. However, the tools and practices described have evident implications in terms of transparency of the processes for the wealth of data about tourism development and evolution in the destination, which creates the potential for knowledge development, and the awareness and engagement of residents and non-tourism stakeholders. This is not an automatic mechanism. This evolution must be conceived and developed by DMOs or public authorities willing to turn these technologies into tools supporting e-democracy and participatory decision-making for tourism policies.

One aspect discussed in the literature that the case analysis confirmed concerns the importance of social capital for creating the conditions for knowledge sharing (Lee 2015; Trunfio and Campana 2019) and effective technology adoption. The case of Goteborg demonstrated how ambitious projects of ICT development for the sustainable destination, aimed at multi-stakeholder engagement, are realistic in those contexts where the history and heritage of networking and collaboration are present. The reputation of the DMO and the collaborative culture rooted in a destination are important preconditions for effective technological investments for the smart and sustainable destination, to reduce the risk of abandonment of the created platforms or limited stakeholder engagement with them.

4.7.3 Emerging Managerial Approaches: Normative Enforcement, Visitor-Oriented Destination Management, and Stakeholder-Oriented Destination Management

The case analysis allows us to discuss the managerial approaches emerging from the ICT projects and practices described. The focus is on the analysed projects and the discussion is limited to the managerial approaches suggested by how these projects are developed and are meant to contribute to a sustainable destination.

The first finding is that *normative enforcement* does not significantly and exhaustively interpret the analysed cases. This means that the adopted technologies are not designed and introduced locally to help the local authority define access limits, regulations, and control over respect for the defined rules. ICT tools may help in this direction, but this does not significantly describe the emerging managerial approaches in the cases analysed, except for Dubrovnik and the Faroe Islands.

In the case of Dubrovnik, the Port Authority defined the maximum number of 8000 cruise tourists daily. Concerning the Faroe Islands and the "Closed for Maintanance" initiative, the technology deployment is minimal and mostly related to the social media and communication strategy. In this case, there is temporally limited access to the islands, and there is room to argue that this configures more as a marketing action. It draws international attention to the Faroe Islands as a sustainability brand, giving voice to the ongoing local commitment to sustainability rather than a normative enforcement approach to managing the sustainable destination.

Visitor-oriented destination management is at the core of several of the ICT tools described. The market-oriented management of the destination, focusing on the market intelligence potential of smart technologies, is prioritised to shape valuable and sustainable tourism products and steer visitors' responsible behaviour (Bramwell et al. 2008; Goodwin 2019; Font and McCabe 2017). For instance, the OnlyLyon Experience, Genoa Exploracity, and the Goteborg Destination Data Platform integrate this managerial approach. The ICT tools allow micro-data collection and knowledge building on actual visitors and forecast trends that help shape valuable and appealing sustainable tourism products and facilitate collaboration across the experience value chain. The responsibility for sustainable behaviour remains with the visitors who have to choose and engage with the local sustainable offering.

Stakeholder-oriented destination management complements the visitor-oriented destination management approach. Besides the centrality of the visitor and his/her choices, some ICT tools rely on the involvement of the different stakeholders (visitors, residents, tourism businesses, and public authorities) in encouraging a sustainability-oriented collective mindset and collaborative action. The OnlyLyon Experience and the Goteborg Destination Data Development are good examples of how the visitor-oriented perspective is complemented by the broader stakeholder perspective, as the commitment of several stakeholders to the functioning of these platforms is a fundamental prerequisite. Co-learning is an evident element of these

ICT tools (Colazzo et al. 2008). If the destination is an evolving system, these platforms help the different stakeholders to learn and achieve a shared vision and collaborative schemes for sustainable product development. The critical element is triggering and orchestrating these complex tools and making them work and grow over time. As has been said in the case of Goteborg, this type of tool can be adopted if the local system has developed a sufficient collaborative capacity amongst local stakeholders.

Forms of stakeholder-oriented destination management are also part of the Lyon City Card, Dubrovnik Smart Parking, and the Dubrovnik Card. These tools target residents and tourists to make these different stakeholders' needs coexist, creating valuable synergies and mechanisms of virtuous self-regulation between the visitors' and residents' demand for public transport, public networks, and a leisure offering.

References

Ali A, Frew AJ (2014a) ICT and sustainable tourism development: an innovative perspective. J Hosp Tour Technol 5:2–16

Ali A, Frew AJ (2014b) Technology innovation and applications in sustainable destination development. Inf Tech Tour 14:265–290

Bekele MK, Town C, Pierdicca R, Frontoni E, Malinverni ES (2018) A survey of augmented, virtual, and mixed reality for cultural heritage. J Comput Cult Heritage 11(2):7–36

Bramwell B, Lane B, McCabe S, Mosedale J, Scarles C (2008) Research perspectives on responsible tourism. J Sustain Tour 16(3):253–257

Buhalis D, Harwood T, Bogicevic V, Viglia G, Beldona S, Hofacker C (2019) Technological disruptions in services: lessons from tourism and hospitality. J Serv Manag 30:484–506

Buhalis D, O'Connor P (2005) Information communication technology revolutionising tourism. Tour Recreat Res 30(3):7–16

Buonincontri P, Micera R (2016) The experience co-creation in smart tourism destinations: a multiple case analysis of European destinations. Inf Tech Tour 16:285–315

Cabiddu F, De Carlo M, Piccoli G (2014) Social media affordances: enabling customer engagement. Ann Tour Res 48(September):175–192

Camero A, Alba E (2019) Smart city and information technology: a review. Cities 93 (October):84–94

Capocchi A, Vallone C, Pierotti M, Amaduzzi A (2019) Overtourism: a literature review to assess implications and future perspectives. Sustain 2019(11):2–18

Cavalheiro MB, Joia LA, Cavalheiro GM (2019) Towards a smart tourism destination development model: promoting environmental, economic, socio-cultural and political values. Tour Plan Develop 17(3):1–23

Colazzo L, Molinari A, Villa N (2008) In: Kendall M, Samways B (eds) From e-learning to ":Co-learning": the role of virtual communities BT–learning to live in the knowledge society. Springer, pp 329–338

Eisenhardt KM (1989) Building theories from case study research. Acad Manag Rev 14(4): 532–550

Eisenhardt K, Graebner M (2007) Theory building from cases: opportunities and challenges. Acad Manag J 50(1):25–32

Errichiello L, Micera R (2017) Special issue of European journal of tourism research smart tourism destinations: advancing theory and practice. Euro J Tour Res 17:5–6

Eurocities (2010) A shared vision on city branding in Europe. Eurocities

Femenia-Serra F, Neuhofer B, Ivars-Baidal JA (2019) Towards a conceptualisation of smart tourists and their role within the smart destination scenario. Serv Ind J 39:109–133

Font X, McCabe S (2017) Sustainability and marketing in tourism: its contexts, paradoxes, approaches, challenges and potential. J Sustain Tour 25(7):869–883

García-Hernández M, Ivars-Baidal J, Mendoza de Miguel S (2019) Overtourism in urban destinations: the myth of smart solutions. Boletin de la Asociacion de Geografos Espanoles 83:1–38

Globetrender (2020) Faroe Islands welcomes thousands of virtual tourists via live video stream. Globetrender. https://globetrender.com/2020/06/07/faroe-islands-virtual-travel-remote-tourism/. Accessed 3 Jan 2023

González AT (2018) Venice: the problem of overtourism and the impact of cruises. Investigaciones Regionales J Reg Res 42:35–51

Goodwin H (2019) Barcelona–crowding out the locals: a model for tourism management? In: Dodds R, Butler RW (eds) Overtourism. Issues, realities and solutions. De Gruyter, Berlin, pp 125–138

Goteborg & Co (2021) Annual and sustainability report. Goteborg & Co. https://goteborgco.se/en/about-goteborgco/business-plan-and-annual-reports/. Accessed 3 Jan 2023

GrandLyon (2021) Schéma de développement du tourisme responsable 2021–2026. GrandLyon. https://business.onlylyon.com/fileadmin/user_upload/documents/20211215-mdl-schema-developpement-tourisme-responsable-metropole-de-lyon-2021-2026-fr.pdf. Accessed 3 Jan 2023

Habtemariam D (2022) Faroe Islands tourism maintains pre-pandemic momentum. Skift. https://skift.com/2022/06/22/faroe-islands-tourism-emerges-from-pandemic-still-immune-to-obscurity/. Accessed 3 Jan 2023

Hjalager AM (2015) 100 innovations that transformed tourism. J Travel Res 54:3–21

Homburg C, Jozic D, Kuehnl C (2017) Customer experience management: toward implementing an evolving marketing concept. J Acad Mark Sci 45:377–401

Ivars-Baidal JA, Celdrán-Bernabeu MA, Mazón JN, Perles-Ivars ÁF (2019) Smart destinations and the evolution of ICTs: a new scenario for destination management? Curr Issue Tour 22(13): 1581–1600

Lam-González YE, León CJ, de León Ledesma J (2019) Assessing the effects of the climatic satisfaction on nautical tourists' onsite activities and expenditure decisions. J Destin Mark Manag 14:100372

Lee BC (2015) The impact of social capital on tourism technology adoption for destination marketing. Curr Issue Tour 18:561–578

Lemon KL, Verhoef PC (2016) Understanding customer experience throughout the customer journey. J Mark 80(6):69–96

Leotta A (2022) Virtual tourism in the age of Covid-19: a case study of The Faroe Islands' 'remote tourism' campaign. In: Bonelli D, Leotta A (eds) Audiovisual tourism promotion. A critical overview. Springer, pp 107–125

Neuhofer B, Buhalis D, Ladkin A (2012) Conceptualising technology enhanced destination experiences. J Destin Mark Manag 1:36–46

Panayiotopoulos A, Pisano C (2019) Overtourism dystopias and socialist utopias: towards an urban armature for Dubrovnik. Tour Plan Develop 16(4):393–410

Pantile D, Novellino A, Verreschi G (2018) Exploracity. An innovative platform offering new services for the enhancement of tourism and cultural heritage, CHNT 2018 Conference.

Schaper MM, Santos M, Malinverni L, Zerbini Berro J, Pares N (2018) Learning about the past through situatedness, embodied exploration and digital augmentation of cultural heritage sites. Int J Hum Comput Stud 114:36–50

Scholz, Agenda F (2020) Compendium of best practices. 2019&2020 European capital of smart tourism competitors. Initiative of the European Union.

Sigala M (2018) New technologies in tourism: From multi-disciplinary to anti-disciplinary advances and trajectories. Tour Manag Perspect 25:151–155

Sigala M, Marinidis D (2012) E-democracy and web 2.0: A framework enabling DMOs to engage stakeholders in collaborative destination management. Tour Anal 17:105–120

Stecker B, Hartmann R (2019) Case study: Balancing the sustainability of tourism in City destinations—The case of dubrovnik, corporate sustainability and responsibility in tourism, pp 373–382

The Faroe Islands (2019) Join the Preservolution! A sustainable tourism development strategy for the Faroe Islands towards 2025. The Faroe Islands. https://visitfaroeislands.com/en/meetings/sustainability/join-the-preservolution. Accessed 3 Jan 2023

Trunfio M, Campana S (2019) Drivers and emerging innovations in knowledge-based destinations: towards a research agenda. J Destin Mark Manag 14:1–12

Trunfio M, Campana S, Magnelli A (2020) Measuring the impact of functional and experiential mixed reality elements on a museum visit. Curr Issue Tour 23(16):1990–2008

Trunfio M, Della Lucia M (2019) Co-creating value in destination management levering on stakeholder engagement value. e-Review Tour Res 16:195–204

Trunfio M, Jung T, Campana S (2022) Mixed reality experiences in museums: exploring the impact of functional elements of the devices on visitors' immersive experiences and post-experience behaviours. Inf Manag 59:1–16

Trunfio M, Pasquinelli C (2021) Smart technologies in the Covid-19 crisis: managing touris flows and shaping visitors' behaviour. Euro J Tour Res 29:2910

Whyte P (2019) Skift global forum preview: how The Faroe Islands got the tourism's world attention. Skift. https://skift.com/2019/08/27/skift-global-forum-preview-how-the-faroe-islands-got-the-tourism-worlds-attention/?utm_campaign=Skift%20Global%20Forum%202019&utm_content=99579440&utm_medium=social&utm_source=facebook&hss_channel=fbp-155942621181219&fbclid=IwAR3S7Tj1AZmFcMh_g30_i6paWFQNSGqfyXf1_lOuPW68SG25sf37wHoDR78. Accessed 3 Jan 2023

Xiao H, Smith LJ (2006) The making of tourism research: insights from a social sciences journal. Ann Tour Res 33(2):490–507

Xu F, Weber J, Buhalis D (2013) Gamification in tourism, Information and communication technologies in tourism, vol 2014, pp 525–537

Yin RK (2014) Case study research: design and methods, 5th edn. Sage, Thousand Oaks, CA

Chapter 5
Framing Sustainability-Oriented Innovation in Smart Tourism Destinations: Conceptual Advancements and Research Agenda

5.1 Sustainability-Oriented Innovation in Smart Tourism Destinations

This book was built on the multifaceted concept of sustainability and a critical analysis of smart technologies' roles alongside sustainable destination development, which represents a field of largely underdeveloped inquiry relevant to academics, practitioners, and policymakers. The "tech-clash" in the contemporary global scenario represents the starting point for this book, which casts a critical light on the consolidated views of disruptive technological exploitation and lays the foundation of the academic debate on sustainable-oriented innovation in smart tourism destinations.

This chapter draws conceptual advancements and sets the research agenda on sustainability-oriented innovation in smart tourism destinations. Sustainability-oriented innovation recombines research on technology- and social-driven innovation in tourism destinations and the sustainable destination agenda, providing an integrative understanding of the nexus between smartness and sustainability within destinations. It addressed the research question of how and to what extent smart technologies drive tourism innovation focused on sustainability and explored the unchartered territories of sustainability-oriented innovation for human wellbeing and societal development. Accordingly, it engaged with an analysis of the role of technology by critically addressing the modalities, opportunities, challenges, and pitfalls of technology deployment in rebalancing tourism disequilibria within local systems. It overcomes the myopic conception of disruptive smart technology deployment and introduces conceptual advancements in the tourism innovation debate, setting a research agenda on destinations' sustainability-oriented innovation.

This integrative perspective of sustainability-oriented innovation in contemporary destinations proposes conceptual advancements by cross-fertilising sustainability, innovation, and smart tourism theoretical frameworks. In so doing, this research connected with the sustainability-oriented innovation (SOI) concept that emerged in

C. Pasquinelli, M. Trunfio, *Sustainability-oriented Innovation in Smart Tourism*, Tourism on the Verge, https://doi.org/10.1007/978-3-031-33677-5_5

the business literature, which focuses on those innovations introducing and integrating social and environmental aspects into products/services, processes, and organisations (Klewitz and Hansen 2014). SOI concerns an organisation's ability to change and reshape products, practices, and internal processes to boost social and environmental value creation (Adams et al. 2016). It highlights an internal change and shift and the transformative effects of business innovations over entire industries, consolidated management practices, and markets (Coenen and Truffer 2012).

The SOI literature has mainly focused on technical and technological aspects of product and process innovation, whilst limited attention was drawn to stakeholder mindsets and cultural meanings (Paramanathan et al. 2004; Adams et al. 2016: Pasquinelli et al. 2023). There is an evident need in the literature to further integrate innovation and sustainability agendas into shared conceptual frameworks working on the overlapping areas and issues and dealing with the dissonant aspects (Pasquinelli and Trunfio 2020). These conceptual advancements are necessary to support innovations oriented towards sustainable value co-creation in tourism destinations.

This book contributed to framing sustainability as an innovation process oriented towards achieving social, economic, and environmental values for destination stakeholders, pursuing the quality of the tourist's and the resident's place experience and human wellbeing. It has explored theoretical domains and identified European destinations, including European Capitals of Smart Tourism, where smart technology deployment accompanies multifaceted transformative processes, engaging stakeholders in sustainable-oriented innovations towards rebalancing tourism disequilibria and fostering inclusiveness and quality of life.

5.2 Sustainability Management and Policy Implications

The destination's SOI conceptualisation opens routes to further analyse how sustainability in the smart destination can be configured as a development strategy. Three points are in focus. First, in light of the essence of the co-evolutionary process, as described above, SOI is likely to develop in the frame of a strategic transition framework (see Chap. 3). As explained in Chap. 3, the strategic framework provides an interpretative opportunity to explain the actions undertaken in the destination (on the analytical side); besides, it gives a sense of orientation to the actions to be taken (from a normative side).

The SOI does not necessarily imply the maintenance of the *status quo* (the conservative framework in Chap. 3 corresponding to the "business as usual" philosophy, Dwyer 2018) for the emerging innovations that may impact the destination, and potentially the hospitality industries and tourism markets. On the other hand, the SOI cannot a priori be framed as a radical transformation strategy (see the radical framework in Chap. 3): we learnt about the progressive and partial changes in the destination system that are intrinsic to the SOI. The transition frames the destination's SOI as a middle-ground strategy developing through incremental steps in the destination system, making sustainability smoothly enter the destination development agenda.

The smart tourism mindset and social capital (Chap. 2) suggest the opportunity to interpret the SOI within the strategic transition framework. Any attempt to define sustainability strategies and consequent actions as either a conservative or transformative risk is not realistic or workable for the destination community. The transition as a middle-ground approach (Hudson 1992, see Chap. 3) suggests the opportunity to implement single actions that, although necessary to trigger SOI, are not directly connected to the "big" sustainability goals. For example, it may be needed to steer small-win projects with local stakeholders, not necessarily aimed at complex sustainable goals, to trigger trust and dialogue. It may be helpful to foster training and education in the local tourism labour market to make workers engaged components of the destination knowledge-based system. These may converge into SOIs in the destination without a well-defined strategic framework.

A second point, closely linked to this first one, concerns the SOI not necessarily developing as a deliberated sustainable destination strategy. As the case analysis in Chap. 4 showed, the SOI can configure as an emerging development strategy. It may emerge over time through previous strategic steps pursuing product and service innovations, technological innovations, and ex post becoming integral to the destination's sustainability goals. This evolution may be due to the local and global environmental and social issues becoming increasingly visible and central for destination development and policy discourse. As mentioned above, actions go in different directions, for instance, designing novel tourism products meant to be more relevant to the market, capturing public funding opportunities, technological advancements, and labour market improvements. They may contribute to consolidating the local co-determinants of the destination's SOI, which might be visualised by the destination community and more systematically pursued only at a later stage.

A third and final point concerns the sustainable tourism vision oriented to and shaped by the SOI process. In connection with the previous point, a well-defined sustainable tourism vision can lead planning for the destination's SOI, or it can emerge from the series of actions that, responding to specific needs and contingencies, are then reframed into a sustainability vision by the destination community in the medium-long term. In this second option, the sustainability vision is integral to the emergence of the smart tourism mindset which, as mentioned, is the constructive and collective re-imagining of local sustainability. As discussed above, it is worth stressing how the definition or emergence of a sustainable tourism vision is important to the SOI for the need to align all the co-determining factors in one direction. The vision must point stakeholders and resources in that same direction. This further highlights the centrality of the orchestrator and sustainable tourism governance, effectively orchestrating the system through pursuing the local sustainability vision and being in charge of aligning factors and stakeholders' individual actions towards its achievement.

The three types of sustainable tourism vision discussed in Chap. 3 were sustainable growth, sustainable development, and degrowth. Future research should investigate the relationship between the sustainable tourism vision and SOI, its development, characteristics, and pitfalls. We learnt that SOI needs to leverage the factors enabling mutual understanding and appreciation of stakeholders' efforts, engagement, and commitment throughout sustainable value creation "with" and "for" *all* the destination stakeholders (see Chap. 3). Appreciation can be mediated by images and perceptions in stakeholders' minds. Research should deepen knowledge of the perceptual domain and its role in SOI (Pasquinelli et al. 2023) within destinations.

Furthermore, the sustainable growth and the sustainable development visions discussed in Chap. 3 align with the SOI conceptualisation. The former combines sustainability with tourism growth as the irrevocable foundation of economically sustainable tourism; the latter refers to the expansion of the qualitative content of tourism products prioritising the improvement of the tourism destination and its offering over quantitative tourism growth.

The degrowth imperative (Higgins-Desbiolles et al. 2019; Cheung and Li 2019), postulating an unavoidable decrease in the quantitative and economic tourism flows towards sustainability, is hard to realistically pursue since engaging fundamental destination stakeholders with this, such as tourism businesses and organisations and tourism workers, may be difficult. These may become economically impacted by this vision or, at least, their perceptions of the consequent negative impacts on their lives, professional activities, and economies may disengage them and induce them not to participate in innovation projects and practices focused on sustainability. This does not mean that the degrowth vision should not be considered realistic or viable; this means that it can hardly be pursued in the frame of sustainability-oriented innovation, which is one specific configuration for pursuing sustainability in the destination.

5.3 Setting the SOI Research Agenda in Smart Tourism Destinations

This research identifies key emerging traits of sustainability-oriented innovation in tourism destinations, opening research routes and offering sustainability management and policy implications.

A set of propositions emerged from the theoretical and empirical explorations developed in this book, and this helps to frame tourism innovation, pursuing sustainable tourism models within destinations. These propositions frame contemporary trends and prospects of evolution in sustainable tourism development in the frame of destination innovation.

5.3.1 *Smart Technologies Boosting the Sustainable Destination as a Place of Inclusion and Wellbeing*

The theoretical debate and the empirical analysis have pointed out a progressive shift from exploiting smart technologies for tourists' experiences to conceiving and designing a sustainability-oriented destination, leveraging the combination of technology- and social-driven innovation. The combination of smart destination, smart business ecosystem, and smart experience as intertwined defining components of smart tourism (Gretzel et al. 2015; Gretzel 2021, see Chap. 2) finds further theoretical and empirical support in this research.

This proposition echoes the consolidating stakeholder perspective that characterises the sustainability agenda, stakeholder-oriented destination management, as defined in this book, and the identification of stakeholder engagement as an analytical dimension of the sustainable destination (see Chap. 3). This shift encompasses a systemic view of the destination as a bundle of experiences meeting all stakeholders' needs, expectations, and preferences, including residents, tourists, businesses, and, generally, city users. This stakeholder perspective is pivotal to the integrative effort of the smartness and sustainability agendas made in this book.

Proposition 1 Smart technologies support not only tourists' experience but, more broadly, residents' and city users' life experiences, contributing to the smart and sustainable destination as a place of inclusion and wellbeing.

5.3.2 *Smart Tourism Mindset Fosters Sustainability-Oriented Innovation*

As the European destinations analysis suggested, smart technologies are increasingly considered an investment in enhancing experiences and assisting multiple stakeholders' mutual engagement with the sustainable destination and stakeholders' benefits (such as experience, accessibility, and inclusiveness). The smart experience can be a lever of sustainability if destination stakeholders actively participate in its design and implementation through co-creative mechanisms. Residents', visitors', and local entrepreneurs' engagement with the sustainable destination implies awareness of the local meaning of sustainability, an active role in local, sustainable development through individual responsible behaviour, and the willingness to participate in local virtuous learning cycles in the local system.

The potential of smart technologies resides in their capacity to make the engagement process and its outcomes and benefits visible and perceived by all the destination stakeholders. Raising visibility and stakeholders' awareness is central to sustainability, the subjective dimension of which was discussed in this book (Chap. 3). Visibility and information sharing about the destination's progress towards sustainability foster stakeholder understanding and appreciation of the efforts made. This is an important mechanism for constantly and progressively engaging stakeholders.

An additional important condition to consider when thinking of stakeholder engagement in technology-driven innovation regards the human capital, skills, competencies, the propensity to use daily, and reliance on technologies. This condition is particularly critical in the tourism industry, where the destination is made up of several small-sized companies. Their contribution—together with the whole local community—to the tourism knowledge ecosystem is fundamental, but barriers to adopting technologies and knowledge absorption can be high. The literature witnessed stakeholders' capabilities and capacity to join ambitious technology-driven innovation projects are essential for building the tourism knowledge ecosystem. Human capital development is an important innovation factor: working on human capital towards a tourism knowledge ecosystem and creating the conditions for nurturing an absorptive capacity in the local system deserve further scholars' attention. These represent challenges aligning with the People pillar of the Sustainable Development Goals, which calls attention to the need for a better tourism labour market. Well-trained and skilled workforce and professional development are the preconditions for inclusive and prosperous tourism development.

Furthermore, destination sustainability-oriented innovation research must draw attention to human–technology interaction, modalities, attitudes, limits, and challenges at the interface between human beings and technologies. The smart tourism mindset concept (Gretzel 2021), based on shared ideas and values amongst stakeholders (see Chap. 2), is in its conceptual infancy. However, it may represent an area for theoretical advancements in the studies on sustainability-oriented innovation in smart destinations.

Proposition 2 The smart tourism mindset of destination stakeholders fosters the co-evolutionary process of constructive and collective re-imagining of sustainability-oriented innovation.

5.3.3 *Knowledge Production, Acquisition, Elaboration, and Big Data*

Smart tourism can capitalise on immersive technologies, artificial intelligence and big data production and exploitation towards sustainability-oriented innovation. Micro-data development and analysis may become the glue of the local knowledge ecosystem as a concrete incentive to engage with the sustainable destination. Referring to the important issue of visibility and appreciation for motivating and maintaining stakeholders' willingness to contribute to the local system, smart technologies that extract, combine, and analyse data help to provide the much-needed visibility and sharing opportunity. Those platforms that collect and analyse data by sourcing them from multiple stakeholders (e.g. visitors and businesses) and providing them with inputs are tools that support decision-making, mutual visibility, and practice sharing. Furthermore, using shared data for measuring impacts and benefits and monitoring their evolution can be a way to raise the visibility of stakeholder behaviour and efforts and, consequently, appreciation of mutual efforts.

However, a condition must be met. Data collection and analysis act as a binding mechanism amongst all the destination stakeholders whenever the system shows the capacity to extract concrete value from micro-data and the connected technological platforms. The individual stakeholders or, as seen in the case analysis, the destination system need the capacity to transform data into operative knowledge and actions. Beyond the adoption of technology allowing big data collection and analysis, there is a need to take care of all those factors and procedures leading to actionable knowledge production. Knowledge should be extracted from data in a suitable shape for the specific destination decision makers (public and private actors).

Proposition 3 Exploration and exploitation of immersive technologies, artificial intelligence and big data drive sustainability-oriented innovation as binding mechanisms boosting engagement and connections amongst destination stakeholders.

5.3.4 Destination Management Organisations

Destination Management Organisations (DMOs) continue to play a relevant role in orchestrating technology-driven innovation towards sustainability that, as stated above, does need a set of co-determining factors beyond technology deployment. The more complex the pursued innovation process and the more advanced the technological investment the destination system undertakes, the more central the role of the DMO. A high level of coordination of a reputable orchestrating organisation allows converging investments in human capital, relationship building, capacity to steer the engagement around innovation processes, raising awareness of sustainability issues and mutual appreciation of all stakeholders' efforts. The "organisational capacity" for sustainability (Rasoolimanesh et al. 2020) is integral to technology-driven innovation processes in the destination.

The previous propositions suggest a sort of technology adoption cycle in the destination system which draws attention to the identification of the "right" technology for different moments in the evolution of the destination system. This temporal dimension certainly deserves further conceptualisation and investigation, as it has important implications. Top-down investments in technology deployment, potentially financing advanced systemic solutions for the sustainable destination, may simply fail and finance a technology that becomes obsolete before exerting any significant impact if the technology is adopted at the wrong moment over the destination lifecycle. The timing is wrong whenever the above-mentioned factors (human capital, stakeholder engagement, social capital, acceptance and propensity to use technology, tourism governance, and organisational capacity) that are participating in triggering technology-driven innovation are not or are inadequately present. This risk characterises several national and supranational funding programmes sustaining tourism innovation embody. The opportunity for funding in the face of the limited investment capacity often characterising local destinations may boost the adoption of advanced solutions. These, however, may be placed in local environments scarcely prepared for their exploitation, fuelling disappointment, frustration, and mistrust amongst destination stakeholders, with potentially deleterious effects on the trajectory of destination development.

The "right" technology for a given destination implies reflecting on the provenance of technology. Earlier in this book, proprietary tailor-made technological solutions were introduced in the empirical exploration. In this regard, sustainable tourism goes with the opportunity to connect tourism with the ICT industry whenever new ad hoc technologies can be usefully developed. This mechanism helps tourism to produce economic effects beyond the conventional tourism boundaries, involving additional actors and making them directly aware of the impacts of tourism. Proprietary technology is not necessarily a promising option for the destination. As mentioned above, proprietary technology is likely an option when the system has developed a clear vision of what is needed and what is the technology-driven innovation process that is likely to boost sustainability in the local context. In this case, the previously discussed factors are even more important as the destination system not only adopts an ICT tool but also conceives and designs it, in collaboration with technological partners, as with the analysed cases witnessed in Chap. 4.

Proposition 4 Destination Management Organisations (DMOs) play a critical orchestrating role in leveraging technology- and social-driven processes towards sustainability-oriented innovation.

5.4 Conceptualising SOI in the Smart Tourism Destination: Emerging Traits and Research Hypotheses

The following statements conceptualise SOI within destinations according to the results from the previous chapters' theoretical and empirical explorations (Table 5.1). These statements represent a first step towards framing destinations' SOI. Further research is needed to develop and test these emerging statements, which should be read as research hypotheses to be investigated and tested. The first statement describes *the destination's SOI as technology- and social-driven innovation*. ICTs are key infrastructures triggering and reinforcing innovations because they allow multiple stakeholders to easily interact, collaborate, and exchange resources (Buhalis 2019; Buhalis and Law 2008; Buhalis and O'Connor 2005; Gretzel et al. 2000; Neuhofer et al. 2012; Racherla et al. 2008; Stamboulis and Skayannis 2003; Trunfio and Campana 2019, 2020). They define actions and solutions that address tourism' impacts and pursue sustainable value creation for the destination stakeholders. Technology supports this process by providing those platforms that, whilst improving and enhancing the tourist experience, assist and

Table 5.1 Conceptualising SOI in the destination: Emerging statements

SOI conceptualisation	Innovation dimension	Sustainability dimension
Technology-driven innovation	ICT infrastructures and tools Interaction, collaboration, exchange	Facilitating actions to address tourism impacts and imbalances
Social-driven innovation	Social capital (norms, values, beliefs, tacit knowledge, trust, and alternative forms of interaction)	Convergence on sustainability values
Smart tourism mindset	Constructive and collective re-imagining	Constructive and collective re-imagining of local sustainability
Knowledge-based social system	Absorptive capacity, propensity to use technology for knowledge prosumers	Mutual understanding and appreciation of stakeholders' efforts, engagement, and commitment
Co-evolutionary process	Technology, smart tourism mindset, and social capital	Alignment towards the local sustainability vision
Orchestrated process	The orchestrator (DMO)	Channelling the bottom-up innovations and innovative forces towards sustainability

facilitate stakeholder coordination and collaboration, as in the case of the micro-data collection and analysis discussed in the previous section. Public funding and investments promote the deployment of technological platforms assisting sustainable development, an opportunity to innovate tourist experiences and the underlying processes related to the need to design competitive bidding projects.

As mentioned, on the other hand, the destination's SOI configures as social-driven innovation. Social capital, as the basis for multiple stakeholders' participation and engagement, plays a crucial role in destination changes and innovation, even though it remains under-investigated (Trunfio and Campana 2020). It corresponds to the bundle of norms, values, beliefs, tacit knowledge, trust, and alternative forms of interaction amongst stakeholders (Coleman 1988; Inkpen and Tsang 2005; Nahapiet and Ghoshal 1998; Putnam 1993).

Research is needed to operationalise and explore social capital as a non-technological platform that co-determines the destination's SOI. Social capital nurtures innovation and facilitates destination development, cultural regeneration, and social inclusion (Go et al. 2013; Trunfio and Campana 2020). Dedicated attention to its role in the SOI mechanism is needed. Furthermore, the intertwining and interactions between social capital and technological platforms deserve attention. The power of relations, relationship building, and the modalities to relate one to the other amongst the destination stakeholders are important aspects of the SOI to investigate.

The nature of the SOI as a technology- and social-driven innovation is further framed by three emerging traits. First, *SOI builds on and contributes to the destination's smart tourism mindset*, representing the bridge connecting the two innovation models discussed in the literature and converging into the SOI. The smart tourism mindset is a constructive and collective re-imagining of changes (Gretzel 2021), the building of which is a situated process characterising a destination community, which collectively translates the abstract and universal notion of sustainability. The notion of sustainability becomes socialised in the destination and progressively takes shape. The emerging mindset further boosts practices of value sharing and the individual stakeholders' perceptions of what is locally necessary for sustainable tourism development.

Second, *SOI builds on and contributes to consolidating the destination knowledge-based social system* (Scott and Laws 2010; Trunfio and Campana 2019; Tuli et al. 2019), which is at the core of value co-creation for the destination stakeholders. The knowledge-based ecosystem is the engine of the destination's SOI as the pivot of technology-driven innovation (stakeholders' absorptive capacity, familiarity, and propensity to technology usage are important innovation factors). It is also the pivot of social-driven innovation as knowledge production, flows, and consumption in the destination support and promote mutual understanding and appreciation of each destination stakeholder' efforts, engagement, and commitment to sustainability. Knowledge production and knowledge flows are fundamental to the destination community to size, interpret, and assess tourism impacts and outline the specific sustainability challenges characterising the destination.

Third, *SOI is a co-evolutionary process* where different factors, such as technology, smart tourism mindset, and social capital, must co-evolve towards a temporal alignment that consolidates the sustainable development trajectory over time. As previously revealed by the empirical exploration and discussed, the SOI emerges throughout the different factors' stages of evolution. Technology complexity and multi-functionality, networking and collaborative capacity, and the collective mindset need to align in terms of sophistication, intensity, and consolidation. The SOI develops through subsequent adjustments towards such an alignment. In contrast, the factors' misalignment may hamper the SOI process, generating negative reactions, frustration, and mistrust amongst the destination stakeholders.

SOI configures an orchestrated process in the destination based on the need for factor alignment. The presence of an orchestrator, whose mission is to frame and implement the sustainable destination (typically the DMO), frames the SOI and determines its likelihood. Although innovation may emerge as decentralised and open dynamics, the capacity of the destination to channel bottom-up innovations and innovative forces towards sustainability is an essential aspect of SOI within the destination.

References

Adams R, Jeanreanaud S, Bessant J, Denyer D, Overy P (2016) Sustainability-oriented innovation: a systematic review. Int J Manag Rev 18(2):180–205

Buhalis D (2019) Technology in tourism-from information communication technologies to eTourism and smart tourism towards ambient intelligence tourism: a perspective article. Tour Rev 75(1):267–272

Buhalis D, O'Connor P (2005) Information communication technology revolutionising tourism. Tour Recreat Res 30:7–16

Buhalis D, Law R (2008) Progress in information technology and tourism management: 20 years on and 10 years after the internet-the state of eTourism research. Tour Manag 29:609–623

Cheung KS, Li LH (2019) Understanding visitor–resident relations in overtourism: developing resilience for sustainable tourism. J Sustain Tour 27(8):1197–1216

Coenen L, Truffer B (2012) Places and spaces of sustainability transitions: geographical contributions to an emerging research and policy field. Eur Plan Stud 20(3):367–374

Coleman JS (1988) Social capital in the creation of human capital. Am J Sociol 94:95–120

Dwyer L (2018) Saluting while the ship sinks: the necessity for tourism paradigm change. J Sustain Tour 26(1):29–48

Go FM, Trunfio M, Della Lucia M (2013) Social capital and governance for sustainable rural development. Stud Agricult Econ 115:104–110

Gretzel U (2021) Conceptualising the smart tourism mindset: Fostering utopian thinking in smart tourism development. J Smart Tour 1(1):3–8

Gretzel U, Sigala M, Xiang Z, Koo C (2015) Smart tourism: Foundations and developments. Electron Mark 25:179–188

Gretzel U, Yuan YL, Fesenmaier DR (2000) Preparing for the new economy: Advertising strategies and change in destination marketing organisations. J Travel Res 39:146–156

Higgins-Desbiolles F, Carnicelli S, Krolikowski C, Wijesinghe G, Boluk K (2019) Degrowing tourism: Rethinking tourism. J Sustain Tour 27(12):1926–1944

Hudson B (1992) Community care planning: Incrementalism to rationalism? Soc Policy Adm 26(3):185–200

Inkpen AC, Tsang EW (2005) Social capital, networks, and knowledge transfer. Acad Manag Rev 30(1):146–165

Klewitz J, Hansen EG (2014) Sustainability-oriented innovation in SMEs: a systemic review. J Clean Prod 65:57–75

Nahapiet J, Ghoshal S (1998) Social capital, intellectual capital, and the organisational advantage: sociological and economic approaches to the analysis of social structure. Acad Manag Rev 23: 242–266

Neuhofer B, Buhalis D, Ladkin A (2012) Conceptualising technology enhanced destination experiences. J Destin Mark Manag 1:36–46

Paramanathan S, Farrukh C, Phaal R (2004) Implementing industrial sustainability: the research issues in technology management. R D Manag 34(5):527–537

Pasquinelli C, Trunfio M (2020) Reframing urban over tourism through the smart-city lens. Cities 102:1–8

Pasquinelli C, Rovai S, Bellini N (2023) Linking place brands and regional innovation: sustainable business strategies leveraging heritage. Reg Stud:1. https://doi.org/10.1080/00343404.2023.2187046

Putnam RD (1993) The prosperous community: social capital and public life. Am Prospect 13:35–42

Racherla P, Hu C, Hyun MY (2008) Exploring the role of innovative technologies in building a knowledge-based destination. Curr Issue Tour 11:407–428

Rasoolimanesh SM, Ramakrishna S, Hall CM, Esfandiar K, Seyfi S (2020) A systematic scoping review of sustainable tourism indicators in relation to the sustainable development goals. J Sustain Tour 1–21:1. https://doi.org/10.1080/09669582.2020.1775621

Scott N, Laws E (2010) Advances in service networks research. Serv Ind J 30(10):1581–1592

Stamboulis Y, Skayannis P (2003) Innovation strategies and technology for experience-based tourism. Tour Manag 24:35–43

Trunfio M, Campana S (2019) Drivers and emerging innovations in knowledge-based destinations: towards a research agenda. J Destin Mark Manag 14:1–12

Trunfio M, Campana S (2020) Innovation in knowledge-based destination: technology-driven versus social-driven. Int J Knowl Based Dev 11(2):76–199

Tuli SC, Hu R, Dare L (2019) Planning a global knowledge city: experience from Melbourne, Australia. Int J Knowl Based Dev 10(1):26–42

The manufacturer's authorised representative in the EU is Springer Nature Customer Service Centre GmbH, Europaplatz 3, 69115 Heidelberg, Germany. If you have any concerns regarding our products, please contact ProductSafety@springernature.com

Printed and bound by CPI Group (UK) Ltd, Croydon, CR0 4YY
01/10/2026
02225917-0002